天然橡胶

前沿热点及其演进的知识图谱分析

◎ 李一萍　王大鹏　主编

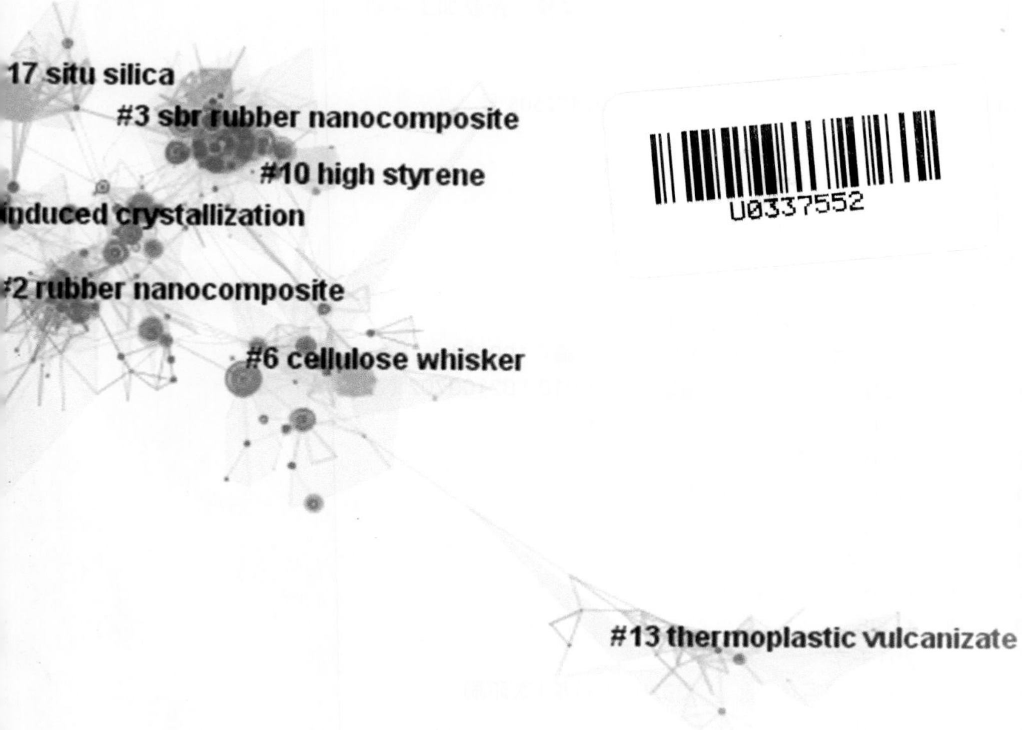

中国农业科学技术出版社

内容简介

本书通过科学计量方法结合CiteSpace、VOSviewer技术，对国内外数据库Web of Science、CSCD和 CNKI收录的天然橡胶科技文献进行文本挖掘和可视化分析，较为全面地展现了天然橡胶领域的研究前沿、研究热点、新兴趋势、科研合作及其演进历程。

本书旨在为中国天然橡胶领域把握国际前沿、引导研究方向和进行产业升级提供科学量化依据，可供天然橡胶化学与材料科学、生物科学以及生态与环境科学等相关专业领域的科技工作者、产业工作者及学生作为专业参考书使用。

图书在版编目（CIP）数据

天然橡胶前沿热点及其演进的知识图谱分析 / 李一萍，王大鹏主编. —北京：中国农业科学技术出版社，2020. 9

ISBN 978-7-5116-5006-1

Ⅰ. ①天… Ⅱ. ①李… ②王… Ⅲ. ①天然橡胶—橡胶加工—研究—中国 Ⅳ. ①TQ332

中国版本图书馆 CIP 数据核字（2020）第 172505 号

责任编辑 李　华　崔改泵
责任校对 贾海霞

出 版 者 中国农业科学技术出版社
北京市中关村南大街12号　　邮编：100081
电　　话 （010）82109708（编辑室）（010）82109702（发行部）
（010）82109709（读者服务部）
传　　真 （010）82106650
网　　址 http: // www.castp.cn
经 销 者 各地新华书店
印 刷 者 北京建宏印刷有限公司
开　　本 710mm × 1 000mm　1/16
印　　张 12.25
字　　数 201千字
版　　次 2020年9月第1版　　2020年9月第1次印刷
定　　价 87.00元

《天然橡胶前沿热点及其演进的知识图谱分析》

编委会

主　编： 李一萍　王大鹏

编　委： 茶正早　邓志声　胡小婵　谢龙莲
李晓娜　曾安逸　曾筱芬

前　言

橡胶是热带地区典型的经济作物，是重要的战略物资。在热带农林业中，橡胶具有特殊和重要地位。迄今为止，天然橡胶在航天、军工及医疗等高端和特殊用途领域中仍具有不可替代性。当前，天然橡胶产业持续低迷，国际天然橡胶产品供大于求，但我国天然橡胶的自给率仍不到20%。长期以来，国内高端和特殊用途的高性能橡胶还几乎完全依赖进口。目前我国天然橡胶产业发展中的一些重大问题已逐渐凸显和转变，如从早期的追求高产转变为高产与优质并重、胶木兼优品种的选育和推广、劳动力成本的不断上升、加工领域工艺改进和技术创新等。与此同时，一个非常值得注意的问题是，在某些高新尖的技术领域正在不断地进行技术革新和突破，如天然高分子或天然橡胶合成纳米复合材料、橡胶微生物降解、橡胶生物合成与调控等，未来可能极大地改变和引领天然橡胶领域发展的新方向。

在天然橡胶产业低迷背景下，追踪国际天然橡胶领域前沿热点，对于我国天然橡胶产业把握国际前沿、引导研究方向和进行产业升级具有重要意义。科学知识图谱是文献计量领域的创新研究方法，能以可视化图形展示科学知识内部的动态演进过程，揭示其研究前沿、研究热点和结构关系等。随着数据科学和计算机科学的快速发展，科学知识图谱的方法也得到了快速发展和传播。诞生于图书情报领域的科学知识图谱在近20年的发展中被广泛地应用于理、工、农、医等各个领域，为展现国际天然橡胶领域前沿热点提供了良好的技术方法。

本书编者们长期从事热带农业文献计量和天然橡胶领域相关的研究工作，对天然橡胶领域的研究前沿、研究热点、新兴趋势、科研合作及其演进历程等进行了深入研究，比较完整地展现了近20年来国际和国内天然橡胶领域的前沿热点和新兴趋势，这些成果可为人们深入了解天然橡胶领域的研究动态、最新动向和发展趋势提供较全面的科学依据。

本书是海南省自然科学基金项目“产业低迷背景下天然橡胶前沿热点追踪研究”（No. 719QN284）和海南省社科基金项目“国际天然橡胶前沿热点及海南橡胶产业升级研究”［No. HNSK（ZC）18-30］的系列成果之一，同时也获得了国家天然橡胶产业技术体系项目（No. CARS-33-ZP-2）、海南省基础与应用基础研究计划项目（2019RC327）和中国热带农业科学院基本科研业务费项目（1630022019007）的大力支持，在此一并致以衷心的感谢。

在本书的撰写过程中，尽管力求详尽、数据准确、全面解读，但限于多种因素的影响，科学、全面和精准地展现天然橡胶研究领域的前沿热点仍存在一定的困难。因此，本书难免存在不妥之处，在此恳请各位读者不吝指正。

编　者

2020年7月

目　录

1 绪论

1.1 引言

天然橡胶是从含橡胶的植物中采割其胶乳加工而成的。据统计，世界上能产胶的植物有2 000多种，其中主要的有大戟科的巴西橡胶树（*Hevea brasiliensis*）、菊科的橡胶草（*Taraxacum brevicorniculatum*）和银色橡胶菊（*Parthenium argentatum*）、杜仲科的杜仲（*Eucommia ulmoides* Oliver）等（国家天然橡胶产业技术体系，2016；International Rubber Research and Development Board，2006）。巴西橡胶树（又称橡胶树）由于其产量高、品质好、经济寿命长、生产成本低等优点，成为人工栽培中最为重要的产胶植物，其产量占世界天然橡胶总产量的99%以上。橡胶是热带地区典型的经济作物，是重要的战略物资。在热带农林业中，橡胶具有特殊和重要的地位。迄今为止，天然橡胶在航天、军工及医疗等高端和特殊用途领域中仍具有不可替代性。当前，天然橡胶产业持续低迷，国际天然橡胶产品供大于求，而我国天然橡胶的自给率不到20%，国内高端和特殊用途的高性能胶几乎完全依赖进口（中华人民共和国国家统计局，2017）。我国天然橡胶产业发展中的一些重大问题已逐渐发生转变，如从早期的追求高产转变为高产与优质并重、胶木兼优品种的选育和推广、劳动力成本的不断上升、加工领域工艺改进和技术创新等。从我国天然橡胶产业发展历程来看，科学技术的提升是推动橡胶产业升级的重要动力。当前一些高新尖的技术领域，如天然高分子或纳米微粒补强天然橡胶合成纳米复合材料、橡胶微生物降解、产胶植物橡胶生物合成与调控等，未来很可能极大地影响天然橡胶产业的发展。

知识图谱是构建科技文献之间连接的研究方法，它通过收集一个知识域的知识集合，反映并呈现科学发展中的新兴趋势和变化。一项知识图谱研究通常包括4个部分：科学文献、可视化分析工具、突出潜在重要模式和趋势的度量指标、揭示可视化知识结构和动态模式的科学理论（Chen，2012a，2017；Chen et al.，2012b；Chen and Leydesdorff，2014；Kim et al.，2016；Ferreira et al.，2016；陈悦等，2015）。随着数据科学和计算机科学的快速发展，科学知识图谱方法得到了快速发展和传播。诞生于图书情报领域的科学知识图谱在近20年的发展中被广泛地应用于理、工、农、医等各个领域（陈超美，2015；Chen，2018）。

1.2 研究目的和意义

挖掘科技文献内部之间的联系并揭示其发展演进规律，已经成为国际文献计量研究的热点领域。中国科学院文献情报中心以共被引分析为基础，每年遴选排名最前的热点研究前沿和新兴研究前沿。而国际上已发展到根据引文分析原理介导的可视化研究阶段，以可视化图谱方式展示科学知识内部的动态演进过程，揭示其研究前沿、演化进程和发展趋势，并广泛地用于多个研究领域。天然橡胶在我国热带农业中仍然具有举足轻重的地位。在我国天然橡胶产业的发展历程中，科学技术对于促进产业发展和产业升级具有重要意义。

作者前期已对国内外天然橡胶文献进行了基本的描述性统计分析，明确了高被引论文、高产机构、学科分布和h指数等文献信息，但还没有深入文献内部进行挖掘，且热带农业计量分析相对薄弱，相关研究也鲜见报道，以天然橡胶为代表的知识图谱分析亟待开展。本书通过科学计量方法结合CiteSpace和VOSviewer技术，对国内外数据库收录的天然橡胶科技文献进行文本挖掘和可视化分析，揭示天然橡胶领域的研究前沿、研究热点、学科前沿及其演进历程等。当今科学技术高速发展，利用科学知识图谱方法对天然橡胶科技文献进行可视化分析，一方面，期望为中国天然橡胶研究跟踪国际前沿、把握研究热点、研究方向及选择科研项目提供一定的科学量化依据；另一方面，对于中国天然橡胶产业升级及政府决策也具有较高的参考价值。

1.3 国内外研究现状

1.3.1 国内天然橡胶产业研究动态

天然橡胶是重要的战略物资。在热带农业中，天然橡胶具有特殊和重要的地位。海南省植胶面积约有850万亩（1亩≈667m^2，全书同），干胶产量接近40万t，是全国两个主产区之一（中华人民共和国国家统计局，2017）。当前天然橡胶价格持续低迷，胶农收入显著下降，生产积极性严重受挫。大多数胶园失管，产胶潜力下降，海南省橡胶产业发展陷于困境。以橡胶种植为主要收入来源的农户，面临着收入低、调整难的困境，目前也尚未找到能够大幅度替代橡胶种植的作物。另外，随着供给侧结构性改革的推进，下游行业对产品质量和多样性的要求也在不断提高。而高端和特殊用途高性能用胶市场几乎被进口天然橡胶抢占。上游技术支撑不足，下游产品低端，天然橡胶产业发展亟待提升（莫业勇等，2017；陈明文，2016；王大鹏等，2013）。在我国天然橡胶产业的发展历程中，科学技术对于促进产业发展和产业升级具有重要意义。一些重大技术革新和应用极大地促进了产业的发展，如一系列高产抗性品种的选育和推广、乙烯利刺激的低频割胶技术等（国家天然橡胶产业技术体系，2016）。

1.3.2 国际天然橡胶研究动态

在1999年以前，国际天然橡胶研究动态集中在天然橡胶生物合成、胶乳过敏、乳胶—水果综合征等方面。Dennis等（1989）证明橡胶延伸因子（REF）是一种与橡胶粒子紧密结合的橡胶粒子蛋白，参与橡胶生物合成中分子链的延伸过程。Oh等（1999）发现小橡胶粒子蛋白（SRPP）可以促进天然橡胶的合成。Turjanmaa（1987）、Kelly等（1993）和Michael等（1996）评估了胶乳过敏的潜在风险因素。Brehler等（1997）证明胶乳交叉反应的IgE抗体能识别胶乳和水果的过敏原。2000—2010年，研究动态以丁苯橡胶或植物纤维素或微晶纤维素/天然橡胶纳米复合材料为主。Ganter等（2001）以丁二烯橡胶和丁苯橡胶为基础，制备有机层状硅酸盐橡胶复合材料。Nair等（2003）从蟹壳中提取纳米微晶纤维素用作天然橡胶的补强填料，制备天然橡胶/纳米微晶纤维素复合材料。Geethamma等（2005）研究椰壳纤维补强天然橡胶及其动态力学性能。Bras等（2010）采用甘蔗渣制得微晶纤维素补强天然橡胶，制备纳米复合薄膜。自2011年开始，出现了纳米微粒补强天然橡胶合成纳米复合材料等方向的研究动态，如对石墨烯进行化学改性，制备高阻隔、高机械性能和高屏蔽性的改性石墨烯/天然橡胶纳米复合材料（Tang et al.，2014；Zhang et al.，2016）。由此可见，国际天然橡胶研究已经形成了新的前沿热点领域，这也是学界亟待追踪和研究的最新前沿热点领域。

1.4 本章小结

从以上研究现状述评可以看出，天然橡胶在热带农业中仍然具有举足轻重的地位，回顾并梳理天然橡胶领域的科技文献，追踪其前沿热点和演变历程，可以为政府部门和海南橡胶产业升级提供决策依据。当今科学技术高速发展，利用知识图谱方法探测天然橡胶领域的国际前沿热点，对于我国天然橡胶产业跟踪国际前沿、把握研究热点及海南橡胶产业升级具有较高的参考价值。

参考文献

陈明文，2016. 我国天然橡胶产业发展形势与因应策略[J]. 农业经济问题，37（10）：91-94.

陈悦，陈超美，刘则渊，等，2015. CiteSpace知识图谱的方法论功能[J]. 科学学研究，33（2）：242-253.

国家天然橡胶产业技术体系，2016. 中国现代农业产业可持续发展战略研究天然橡胶分册[M]. 北京：中国农业出版社. 11-35.

莫业勇，杨琳，2017. 2016年国内外天然橡胶生产形势[J]. 中国热带农业（2）：20-22.

王大鹏，王秀全，成镜，等，2013. 海南植胶区天然橡胶产量提升的问题及对策[J]. 热带农业科学，33（6）：66-70.

中国科学院科技战略咨询研究院，中国科学院文献情报中心，科睿唯安. 2017研究前沿[R/OL].（2017-11-13）[2018-3-5]. https://clarivate.com.cn/eclarivate/pdf/2017_research.pdf.

中华人民共和国国家统计局，2017. 中国统计年鉴[M/OL]. 北京：中国统计出版社，2017[2018-4-25]. http://www.stats.gov.cn/tjsj/ndsj/2017/indexch.htm.

Bras J，Hassan M L，Bruzesse C，et al.，2010. Mechanical，barrier，and biodegradability properties of bagasse cellulose whiskers reinforced natural rubber nanocomposites[J]. Industrial Crops and Products，32（3）：627–633.

Brehler R，Theissen U，Mohr C，et al.，1997. “Latex-fruit syndrome”：frequency of cross-reacting IgE antibodies[J]. Allergy，52（4）：404–410.

Chen C M，2012a. Predictive effects of structural variation on citation counts[J]. Journal of the Association for Information Science and Technology，63（3）：431–449.

Chen C M，2017. Science mapping：a systematic review of the literature[J]. Journal of Data and Information Science，2（2）：1–40.

Chen C M，Hu Z，Liu S B，et al.，2012b. Emerging trends in regenerative medicine：a scientometric analysis in CiteSpace[J]. Expert Opinion on Biological Therapy，12（5）：593–608.

Chen C M，Leydesdorff L，2014. Patterns of connections and movements in dual-map overlays：a new method of publication portfolio analysis[J]. Journal of the Association for Information Science and Technology，65（2）：334–351.

Dennis M S，Light D R，1989. Rubber elongation factor from *Hevea brasiliensis*. Identification，characterization，and role in rubber biosynthesis[J]. Journal of Biological Chemistry，264（31）：18 608–18 617.

Ferreira J J M，Fernandes C I，Ratten V，2016. A co-citation bibliometric analysis of strategic management research[J]. Scientometrics，109（1）：1–32.

Ganter M，Gronski W，Reichert P，et al.，2001. Rubber nanocomposites：morphology and mechanical properties of BR and SBR vulcanizates reinforced by organophilic layered silicates[J]. Rubber Chemistry and Technology，74（2）：221–235.

Geethamma V G，Kalaprasad G，Groeninckx G，et al.，2005. Dynamic mechanical behavior of short coir fiber reinforced natural rubber composites[J]. Composites Part A：Applied Science and Manufacturing，36（11）：1 499–1 506.

Gopalan Nair K，Dufresne A，2003. Crab shell chitin whisker reinforced natural rubber nanocomposites. 2. Mechanical behavior[J]. Biomacromolecules，4（3）：666–674.

International Rubber Research and Development Board，2006. Portrait of the global rubber industry[M]. Kuala Lumpur：IRRDB，73–86.

Kelly K J，Kurup V，Zacharisen M，et al.，1993. Skin and serologic testing in the diagnosis of latex allergy[J]. Journal of Allergy and Clinical Immunology，91（6）：1 140–1 145.

Kim H J，Jeong Y K，Song M，2016. Content-and proximity-based author co-citation analysis using citation sentences[J]. Journal of Informetrics，10（4）：954–966.

Michael T，Niggemann B，Moers A，et al.，1996. Risk factors for latex allergy in patients with spina bifida[J]. Clinical & Experimental Allergy，26（8）：934-939.

Tang Z H，Zhang L Q，Feng W J，et al.，2014. Rational design of graphene surface chemistry for high-performance rubber/graphene composites[J]. Macromolecules，47（24）：8 663-8 673.

Turjanmaa K，1987. Incidence of immediate allergy to latex gloves in hospital personnel[J]. Contact Dermatitis，17（5）：270-275.

Zhang X，Wang J，Jia H，et al.，2016. Multifunctional nanocomposites between natural rubber and polyvinyl pyrrolidone modified graphene[J]. Composites Part B：Engineering，84：121-129.

2 科学知识图谱及其应用

2.1 引言

当前，我们已经进入科研大数据时代，面对海量的科研数据如何才能高效的识别重要的研究成果，特别是梳理研究的发展脉络和研究趋势，成为每一位科研人员关注的问题。在此背景下，科学知识图谱的技术和方法成为解决此类问题的一种可能途径。科学知识图谱（学术地图）是在科学计量与文献计量的理论与方法的支撑下，通过对科技文本的可视化的技术来展示研究主题知识分布、结构和关联的新兴研究方向（陈超美，2015；李杰，2017，2018）。随着数据科学和计算机科学的快速发展，科学知识图谱的方法也得到了快速发展和传播。诞生于图书情报领域的科学知识图谱在近20年的发展中已经扩散到理、工、农、医的各个领域（陈超美，2015；Chen，2018）。与此同时，科学知识图谱领域的分析工具也层出不穷，极大地降低了不同领域学者绘制科学知识图谱的门槛。不同的研究工具，数据可视化设计功能上各有优劣。本书采用的主要知识图谱工具包含CiteSpace和VOSviewer，以使得绘制的科学知识图谱可视化效果准确清晰。

2.2 科学文献引文的价值

1964年，Garfield E首次提出了科学引文索引（SCI）（Garfield E et al.，1964），并利用文献间的引用关系构建了某研究领域的引文网络（Citation Network）（Hummon and Dereian，1989）。文献之间的引用关系主要有：直接引用（Direct Citation）（Albert et al.，1991）、文献耦合（Bibliographic

Coupling）（Kessler，1963）和共被引（Co-Citation）（Small，1973；Marshakova-Shaikevich，1973）。文献之间3种类型的引用关系既可以反映知识的扩散方向，又可以识别某研究领域的知识源头，进而构成原始的科技创新演化路径。随着时间的推移，各类科技文献数量不断增加、知识不断丰富，引文关系网络逐渐复杂，形成了许多需要进一步深入挖掘研究的大样本数据引文网络（廖君华等，2019）。为了有效判别某领域内科技创新的发展动向，科研工作者希望借助复杂的引文关系网络来预测科技创新热点、前沿和发展趋势。

科学文献提供了大量的信息。科学文献中的引文体现了专家学者们对现有文献的选择。无论这种选择是出于何种动机及其具体原因，选择本身提供的信息就很有价值。科学文献可以大致分为经典文献、昙花一现的文献和里程碑式文献3类。经典文献的定义广泛，只要一篇论文不断地被引用，那它就属于经典文献，并非只有爱因斯坦的论文才能成为经典文献。昙花一现的文献占据了科学文献整体中的绝大多数，他们的出现几乎立刻被学者们所遗忘，甚至根本没有引起任何人的关注。第三类文献往往是问题的关键，这些文献从茫茫论文的海洋中产生了飞跃，给人们对于科学知识的认识过程留下了明确的印记。引文分析虽然有弱点和不足，但是它所研究的信息是难以替代的（Chen，2017；Chen et al.，2014；陈超美，2015）。从学者的阐述论证中会学到很多，而从学术同仁对其优劣的描述和评判中能学到更多更深刻的认识。更重要的是，可以明晰学术同仁作出其评判时所依据的逻辑推理和演绎过程。如果能把来自不同学派和不同视角的这种学术鉴定予以综合归纳，那将会极大地减少专家撰写的系统综述中在所难免的个人偏见。这里所指的个人偏见没有任何贬义，这是人类认识、兴趣、经验和观念的必然结果。

2.3 科学计量学与科学知识图谱的产生背景

文献计量学最早是对各种出版发行的期刊或书籍进行测量，早期图书管理员用这种方法找出核心期刊，改善图书馆藏书，观察使用趋势，作为藏书决策的科学依据。科学计量学诞生于20世纪60年代初，是一门用数学方法研究科学的量化特征和发展机制的学科。科学计量学通过对大量有价值的信息进行有效、定量分析，获取与存储、知识的生产与流动、知识的离散与重组，辅助学科领域的研究者发现领域研究的热点及前沿，并作出科学评价（许海云等，2018；许海云和方曙，2013）。科学计量学是将文献计量学应用在科学学的探索中，科学计量学的研究比图书管理员分析科学期刊更为丰富，其中包含了研究表现、创新趋势、科学沟通、领域结构与动态以及政策相关的项目，如基金分析等（陈超美，2015；Chen，2018）。科学知识图谱（学术地图）是以知识领域为研究对象，以显示知识的发展进程与结构关系为目的的一种图形（Shiffrin and Borner，2004）。知识图谱通过将应用数学、信息科学、计算机科学等学科的理论与方法结合文献计量中的引文分析、共现分析等方法，并利用可视化图谱形象地展示学科的核心结构、发展历史、前沿领域及知识架构，以达到多学科融合的现代理论（图2-1）。

INTRODUCTION

Mapping knowledge domains

Richard M. Shiffrin and Katy Borner

PNAS April 6, 2004 101 (suppl 1) 5183-5185; https://doi.org/10.1073/pnas.0307852100

Article　Info & Metrics　PDF

The term "mapping knowledge domains" was chosen to describe a newly evolving interdisciplinary area of science aimed at the process of charting, mining, analyzing, sorting, enabling navigation of, and displaying knowledge. This field is aimed at easing information access, making evident the structure of knowledge, and allowing seekers of knowledge to succeed in their endeavors. Although thousands of years old, this area has undergone a sea change in the last 15 years, a change fostered by an explosion of the amount of information available, the accessibility of that information due to electronic storage, and the new techniques of analysis, retrieval, and visualization that are made possible by vast increases in computational storage capacity and processing speed and power. Many of us are so involved in the new ways of accessing knowledge that we have forgotten how recent is the change to computerized knowledge retrieval with search engines operating on the World Wide Web. Remarkable as these changes are to date, they are only a hint of the transformation to come. The Arthur M. Sackler Colloquium on Mapping Knowledge Domains, held at the Arnold and Mabel Beckman Center of the National Academies of Sciences and Engineering in Irvine, CA, May 9-11, 2003, was designed to showcase the ongoing developments in this transformation and provide pointers toward the directions it will move.

The changes that are taking place profoundly affect the way we access and use information. Scientists, academics, and librarians have historically worked hard to codify, classify, and organize knowledge, thereby making it useful and accessible. The day is fast approaching when all this knowledge will be coded electronically, but mixed in a vast and largely disorganized and often unreliable sea of mostly recent information. Fishing this sea for

Article Alerts　Share
Email Article
Citation Tools　Mendeley
© Request Permissions

PNAS

Table of Contents
Submit

Sign up for Article Alerts
Enter Email Address　Sign up

JUMP TO SECTION
Article
Overview of Contributions
Opportunities and Challenges
Acknowledgments
Footnotes
Info & Metrics

图2-1　科学领域图谱论文（Shiffrin and Borner，2004）

Fig. 2-1　The paper of mapping knowledge domains（Shiffrin and Borner，2004）

2003年5月美国国家科学院组织的一次研讨会，当时会议的组织和参与者包含了最为知名的多个关于科学计量和数据可视化专家学者，如史蒂夫·莫利斯（Steven Morris）、陈超美（Chaomei Chen）、尤金·加菲尔德（Eugene Garfield）以及凯蒂·纳博（Katy Börner）等。会议的议题包含了数据库、数据格式与存取，数据分析算法，可视化与交互设计，应用前景，共4个部分。2004年4月大连理工大学刘则渊教授受到《参考消息》中一篇题为《科学家拟绘制科学门类图》的启发，在国内首先带领自己的团队开始了“科学知识图谱”研究工作，并相继创建了WISE实验室，为我国培养了一批专门从事科学知识图谱实践研究的专业人才。

2.4　科学知识图谱的研究范畴和作用

科学知识图谱研究涉及一个学术领域发展变化的宏观规律及从宏观到微观的变化规律，其宏观规律源于以下3个理论基础：一是Thomas Kuhn的科学革命结构的哲学理论，科学的发展体现为科学范式间的竞争及其转换。二是Stephen Fuchs的社会学思想，对新发现的追求源于对学术名誉威望和资源的竞争。三是Shneider的学科发展理论，即经历了4个阶段，引入新概念和新视角，构造新工具，应用新工具（发现新问题，推广到其他领域），整理积累前几个阶段的知识（Kuhn and Hawkins，1963；Fuchs，1993；Shneider，2009）（图2-2）。从宏观到微观表现在“海尔迈耶问题”（Heilmeier Catechism），一项科学研究想要解决的问题，解决这个问题的重要性及为什么重要，针对这个问题前辈和同行的研究进展，这个问题至今未解决的原因，我们现在所能做到的而前人或他人至今仍未做到；进一步引申为：一是分析对象（领域、学科、专题、个人和团体）。二是问题的模式和研究现状（领域、学科和专题的最终目的和现状）（Chen et al.，2010；Chen，2017，2018）。总的来说，就是对一个学科领域进行系统文献综述。科学知识图谱可以对一个学科领域进行系统的文献综述，其目的是回答一个学科领域的研究历史、里程碑和转折点、主要演进路径，以及这个领域的研究现状、新兴趋势、未来发展方向、尚未解决问题、研究空白点等。

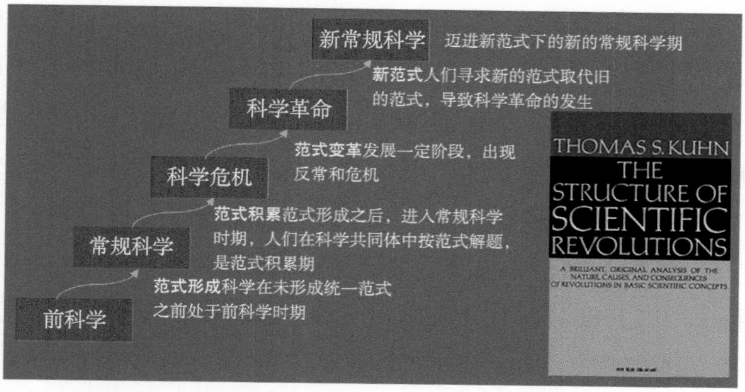

图2–2 库恩的科学发展模式理论（Kuhn T S and Hawkins D，1963）

Fig. 2–2 Kuhn's structure of scientific revolutions（Kuhn T S and Hawkins D，1963）

2.5 科学知识图谱的应用

科学计量学通过定量分析科学文献中的模式来识别研究领域的新兴趋势。文献的可视化分析提供了一种有价值的、及时的、可重复和灵活的方法，除了用于传统的文献综述之外，还可以用于跟踪新趋势的发展并识别关键证据（Chen，2017；Chen et al.，2014；陈超美，2015）。一系列知识图谱工具广泛用于科学计量研究，如HistCite、VOSviewer、Network WorkBench和CiteSpace等（Garfield，2004；van and Waltman，2009；Börner et al.，2010；Chen et al.，2014）。CiteSpace是累加式知识域分析工具，用于对科学领域中文献的新兴模式和重要变革进行可视化分析（Chen，2017；Chen et al.，2014；陈超美，2015）。在科学计量领域，科学知识图谱是可视化分析和领域分析的主要研究手段，它通过收集一个科学知识领域的知识集合，反映并呈现科学知识整体发展过程中的信息。一个研究领域的发展过程可能经历恒定的和革命性的变化，科学知识图谱的一个重要作用就是能够突出潜在的、重要的变化模式和趋势，帮助我们探索和解读可视化的知识结构和动态模式（Chen，2017；Chen et al.，2014；陈超美，2015）。

知识图谱方法是通过构建论文检索式从科技论文数据库中采集研究论文，并使用科学计量及知识图谱的方法对研究论文进行内容分析，在自然科学和社会科学中都有应用，涉及生态学（项国鹏等，2016）、园艺学（刘彬等，2015）、土壤学（吴同亮等，2017）、地理学（桂钦昌等，2016）、经济学（张露等，2016）、环境科学（朱宇恩等，2017）、食品科学（郑江平等，2019）、安全科学（Li and Hale，2016）、管理科学（胡洋等，2017；施萧萧等，2017）、信息科学（程结晶等，2017）和医学（Chen et al.，2012）等多个学科领域（图2-3）。例如，梳理国内外工业生态学研究成果，挖掘工业生态学的前沿热点、知识基础和发展脉络（项国鹏等，2016）；探测土壤科学主题文献的研究热点和主题领域（吴同亮等，2017）；系统回顾1968—2016年Web of Science数据库信息科学领域的新兴趋势和变化等（程结晶等，2017）；解析CSSCI收录的中国政府绩效管理文献的研究热点与前沿（包国宪等，2016）；分析食品科学领域的国际合作规模等（郑江平等，2019）。

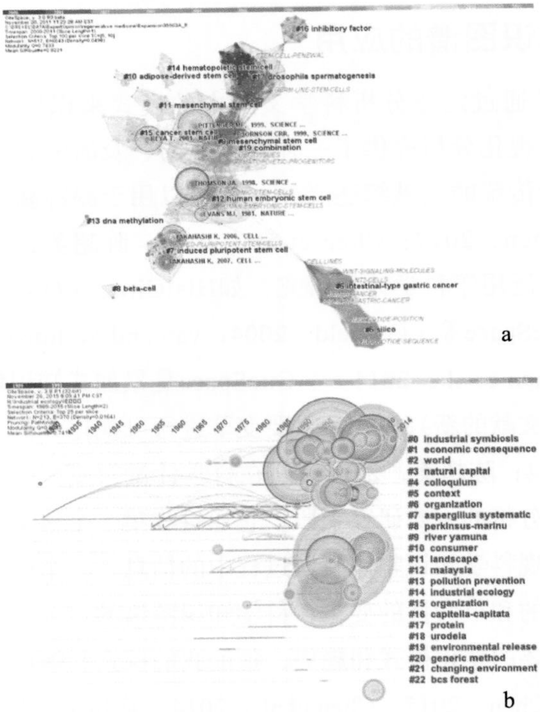

图2-3　再生医学文献共被引和突发主题混合网络图谱（a）；工业生态学研究文献时间线图谱（b）（Chen et al.，2012；项国鹏等，2016）

Fig. 2-3　A hybrid network of co-cited references and burst terms in regenerative medicine（a）; A timeline-view of cited references in industrial ecology research（b）（Chen et al.，2012; Xiang et al.，2016）

2.6 本章小结

随着信息技术的飞速进步，可视化应用越来越普及，今后的学习越来越多地借助各种可视化手段。运用自然语言处理技术和可视化技术，对科学文献进行一系列准确、高效的分析和处理，以求透视出文本或引文关系中蕴含的具有科学价值的内容。科学知识图谱是以表达传递信息、再现过程、找出原因等目标为主的可视化分析的过程或结果。科学知识图谱是为了分析和展示一个学术领域在宏观和微观上的结构、动态特征及发展趋势。科学知识图谱与传统的系统综述异曲同工，相辅相成。科学知识图谱又有独特的优势，不仅可以追溯历史，跟踪当前，还有潜力探索无人涉猎的路径。

参考文献

包国宪，向林科，2016.中国政府绩效管理知识图谱分析[J]. 兰州大学学报（社会科学版），44（2）：46–53.

陈超美，2015. 转折点创造性的本质[M]. 北京：科学出版社. 106–130.

程结晶，丁慢慢，朱彦君，2017. 国外信息管理领域知识流的新兴趋势及可视化分析[J]. 现代情报，37（4）：170–177.

桂钦昌，刘承良，董璐瑶，等，2016. 国外交通地理学研究的知识图谱与进展[J]. 人文地理，31（6）：10–18.

胡洋，赵又霖，2017. 国际信息管理学科领域发展的分析与探讨[J]. 情报科学，35（5）：165–170.

李杰，2017. 科学计量与知识网络分析[M]. 北京：首都经济贸易大学出版社. 332–337.

李杰，2018. 科学知识图谱原理及应用[M]. 北京：高等教育出版社. 16–17.

廖君华，陈军营，白如江，2019. 基于引文内容的多维度科技创新路径构建与可视化研究[J]. 山东理工大学学报（社会科学版），35（4）：80–90.

刘彬，邓秀新，2015. 基于文献计量的园艺学基础研究发展状况分析[J]. 中国农业科学，48（17）：3 504–3 514.

施萧萧，张庆普，2017. 基于共词分析的国外颠覆性创新研究现状及发展趋势[J]. 情报学报，36（7）：748–759.

吴同亮，王玉军，陈怀满，等，2017. 基于文献计量学分析2016年环境土壤学研究热点[J]. 农业环境科学学报，36（2）：205–215.

项国鹏，宁鹏，黄玮，等，2016. 工业生态学研究足迹迁移——基于CiteSpace Ⅱ的分析[J]. 生态学报，36（22）：7 168–7 178.

许海云，董坤，隗玲，等，2018. 科学计量中多源数据融合方法研究述评[J]. 情报学报，37（3）：318-328.

许海云，方曙，2013. 科学计量学的研究主题与发展——基于普赖斯奖得主的扩展作者共现分析[J]. 情报学报，32（1）：58-67.

张露，张越，张俊飚，等，2016. 农业经济管理学科领域的研究发展：历史与前沿[J]. 华中农业大学学报（社会科学版）（3）：31-38.

郑江平，傅天珍，叶兴乾，等，2019. 食品科学领域国际合作论文的文献计量分析[J]. 中国食品学报，19（7）：311-318.

朱宇恩，张倩茹，张维荣，等，2017. 基于文献计量的Cr污染土壤修复发展历程剖析（2001—2015年）[J]. 农业环境科学学报，36（3）：409-419.

Albert M B，Avery D，Narin F，et al.，1991. Direct validation of citation counts as indicators of industrially important patents[J]. Research Policy，20（3）：251-259.

Börner K，Huang W，Linnemeier M，et al.，2010. Rete-netzwerk-red：analyzing and visualizing scholarly networks using the Network Workbench Tool[J]. Scientometrics，83（3）：863-876.

Chen C M，2017. Science mapping：a systematic review of the literature[J]. Journal of Data and Information Science，2（2）：1-40.

Chen C M，2018. Cascading citation expansion[J]. Journal of Information Science Theory and Practice，6（2）：6-23.

Chen C M，Dubin R，Kim M C，2014. Emerging trends and new developments in regenerative medicine：a scientometric update（2000-2014）[J]. Expert Opinion on Biological Therapy，14（9）：1 295-1 317.

Chen C M，Hu Z，Liu S，et al.，2012. Emerging trends in regenerative medicine：a scientometric analysis in CiteSpace[J]. Expert Opinion on Biological Therapy，12（5）：593-608.

Chen C M，Ibekwesanjuan F，Hou J，et al.，2010. The structure and dynamics of co-citation clusters：a multiple perspective co-citation analysis[J]. Journal of the Association for Information Science and Technology，61（7）：1 386-1 409.

Fuchs S，1993. A sociological theory of scientific change[J]. Social Forces，71（4）：933–953.

Garfield E，2004. Historiographic mapping of knowledge domains literature[J]. Journal of Information Science，30（2）：119-145.

Garfield E，Sher I H，Torpie R J，1964. The use of citation data in writing the history of science[M]. Philadelphia，PA：Institute for Scientific Information，187.

Hummon N P，Dereian P，1989. Connectivity in a citation network：the development of DNA theory[J]. Social Networks，11（1）：39–63.

Kessler M M，1963. Bibliographic coupling between scientific papers[J]. American Documentation，14（1）：10–25.

Kuhn T S，Hawkins D，1963. The structure of scientific revolutions[J]. American Journal of Physics，31（7）：554–555.

Li J，Hale A，2016. Output distributions and topic maps of safety related journals[J]. Safety Science，236–244.

Marshakova-Shaikevich I，1973. System of documentation connections based on references[J]. Nauchno Tekhnicheskaya Informatsiya Seriya，2：3–8.

Shiffrin R M，Borner K，2004. Mapping knowledge domains[J]. Proceedings of the National Academy of Sciences of the United States of America，101：5 183–5 185.

Shneider A M，2009. Four stages of a scientific discipline；four types of scientists[J]. Trends in Biochemical Sciences，34（5）：217–223.

Small H，1973. Co-citation in the scientific literature：a new measure of the relationship between two documents[J]. Journal of the American Society for Information Science，24（4）：265–269.

van Eck N，Waltman L，2009. Software survey：VOSviewer，a computer program for bibliometric mapping[J]. Scientometrics，84（2）：523–538.

3 国际天然橡胶文献计量分析

3.1 引言

橡胶是热带地区典型的经济作物，是重要的战略物资（中国热带农业科学院，2015）。国内外科研人员围绕天然橡胶开展了大量研究工作，取得了重要研究进展，SCI收录天然橡胶研究文献也在逐年增加。开展天然橡胶科技论文统计分析工作，有助于对天然橡胶及其相关学科的发展过程、研究现状、研究能力进行科学评价。文献计量学是信息科学广泛采用的一种研究方法，对于计量和分析科技文献收录检索数据具有统计学意义上的合理性和可信度（De Winter et al.，2014；Cañas-Guerrero et al.，2013）。应用文献计量的方法统计和分析科技文献结果，是评价研究成果的必要参考指标之一（Bornmann and Mutz，2015）。SCI即《科学引文索引》（Science Citation Index），是由美国科技信息研究所（Institute for Scientific information，简称ISI）创建的著名的引文索引数据库，是利用文献计量评价科研成果的重要分析工具和文献数据来源（Ho and Kahn，2014）。国内外农业领域的文献计量研究较多，如刘彬等将文献计量学用于国内外园艺植物发展态势的研究中。孙秀焕和路文如对Web of Science收录的水稻文献进行统计分析。Bartol和Mackiewicz-Talarczyk评价了Google学术、Web of Science等收录的纤维作物文献的发展趋势。Von Mark和Dierig对8种新的油料作物的相关文献进行计量研究（刘彬和邓秀新，2015；孙秀焕和路文如，2012；Bartol and Mackiewicz-Talarczyk，2015；Von Mark and Dierig，2012）。在热带农业中，天然橡胶具有特殊和重要的地位。热带作物文献计量研究相对薄弱，以橡胶树为代表的热带作物文献计量研究亟待开展。本章采用文献计量学的方法对1996—2015年SCI收录的天然橡胶科技论文进行统计分析，综述天然橡胶基础和应用研究的发展状况。通过对天然橡胶SCI科技论文进行统计分析，可以为热带科研事业发展定位及相关决策提供科学依据。

3.2 数据来源与研究方法

3.2.1 数据来源

本章所涉及的SCI文献数据均来源于SCI-E数据库中的检索结果。Web of Science（WoS）是Thomson Reuters（汤森路透）集团基于Web而构建整合的数字研究环境。Web of Science数据库收录了23 000多种世界权威的、高影响力的学术期刊，内容涵盖自然科学、工程技术、生物医学、社会科学、艺术与人文等领域，包含Science Citation Index Expanded（SCIE，科学引文索引）、Social Sciences Citation Index（SSCI，社会科学引文索引）和Arts & Humanities Citation Index（A&HCI，人文与艺术引文索引）等7种综合性引文数据库（Goodwin，2014）。

发文量：检索到的某一范围内的论文数量。

引文量：论文所引用的参考文献的数量，是论文的重要组成部分。

篇均他引次数：检索到的某一范围内的所有文献去除自引的被引频次除以发文量。

h指数（H-index）：对于一个国家（地区）的h指数是指，在一定时间内发表的论文至少有h篇的被引频次不低于h次，同时要满足h为最大（Waltman and Van Eck，2012；Hirsch and Buela-Casal，2014）。

影响因子（Impact Factor）：期刊在某年的影响因子是该期刊前两年发表的论文在统计当年被引用总次数除以该期刊前两年发表的论文数（Finardi，2013）。

高被引论文（Highly Cited Paper）：指同一年同一个ESI学科发表论文的被引用次数按照由高到低进行排序，排在前1%的论文（Miyairi and Chang，2012；Knudson，2014）。

3.2.2 研究方法

利用Web of Science平台，选择Science Citation Index Expanded（SCI-E）

引文数据库，采用检索式TI=（“rubber tree*” or “rubber plantation” or “rubber forest” or “natur* rubber” or “rubber latex*” or “rubber tapping” or “*Hevea*”）检索出天然橡胶研究的科技论文，选择文献类型为Article，语种为English进行精炼。时间跨度为1996—2015，检索日期为2016年2月1日。然后对文献记录数、国家/地区、研究机构、发文期刊、影响因子、研究方向、篇均他引次数、h指数和进入ESI前1%学科论文数等进行统计分析。

3.3 国际天然橡胶论文的总体概况

1996—2015年，天然橡胶研究被SCI收录文献共计3 840篇，其中，3 273篇文献被引用，被引频次总计42 819次，去除自引的被引频次总计29 105次，施引文献19 395篇，去除自引的施引文献16 583篇，每篇文献平均引用次数11.15次，单篇被引次数最高为371次，h指数为71；文献数量逐年增加（2001年和2004年除外），引文量也逐年递增。1996—2005年，天然橡胶研究SCI论文发文量较少，共1 294篇，占论文总数量的33.70%，引文量也较少，共6 272篇。自2006年，天然橡胶研究的SCI论文发文量大幅提升，至2015年总计2 546篇，占论文总数量的66.30%，引文量达35 787篇（图3-1）。发文量的增加说明了天然橡胶研究的关注程度和研究实力在不断提升，引文量的增加体现了天然橡胶研究文献参考已有成果的丰富程度。

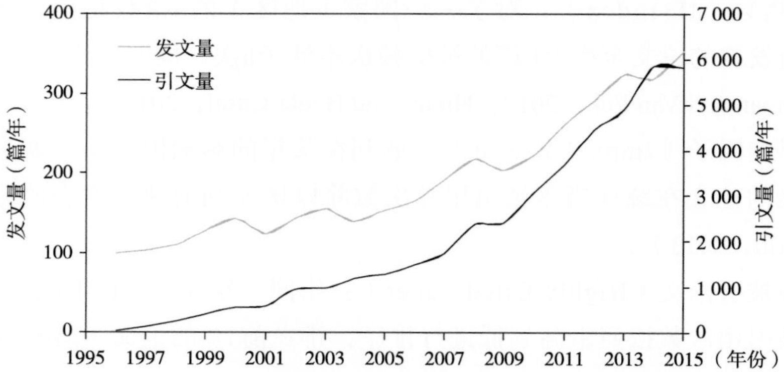

图3-1　1996—2015年天然橡胶SCI论文发文量和引文量

Fig. 3-1　The number and citation amount of nature rubber papers published in SCI-indexed journals during 1996-2015

3.4 国际天然橡胶论文的主要国家/地区以及主产国论文情况

从文献分布的国家/地区来看，1996—2015年，天然橡胶研究被SCI收录文献涉及的国家/地区共91个，主要产胶国泰国、马来西亚、印度和中国SCI论文数量排名世界前4位，发文量占论文总数的60.01%；德国、美国、法国和日本文献的篇均他引次数排在前4位。1996—2005年，在天然橡胶文献数量排名前10位的国家中，中国文献的h指数排名第9位，篇均他引次数排名第6位，2006—2015年，h指数排名上升到第4位，篇均他引次数排名下降到第9位（表3-1）。说明关注这一研究领域的是天然橡胶主产国和工农业发达的国家，主要产胶国SCI论文的数量排在世界前列，发达国家SCI论文的学术水平和影响力都较高，中国文献的学术水平和影响力还有待进一步提高。

表3-1 1996—2015年天然橡胶SCI收录文献数排名前10位的国家

Tab. 3-1 The top 10 most productive countries of nature rubber papers published in SCI-indexed journals during 1996-2015

	1996—2015					2011—2015				
排序 Rank	国家/地区 Country/ Region	记录数 Record count	百分比（%） Percent（%）	h指数 H-index	篇均他引次数 Average cited times without self-citations	国家/地区 Country/ Region	记录数 Record count	百分比（%） Percent（%）	h指数 H-index	篇均他引次数 Average cited times without self-citations
1	泰国	655	17.06	33	7.16	中国	324	20.94	16	2.90
2	马来西亚	610	15.89	38	8.69	泰国	308	19.91	16	3.07
3	印度	536	13.96	32	9.79	马来西亚	292	18.88	13	2.74
4	中国	503	13.10	28	6.19	印度	151	9.76	12	2.89
5	法国	348	9.06	36	14.21	法国	137	8.86	16	5.67
6	美国	341	8.88	40	16.07	美国	105	6.79	13	5.03
7	日本	266	6.93	32	12.11	巴西	104	6.72	10	3.07
8	巴西	238	6.20	20	6.84	日本	74	4.78	9	2.64
9	德国	165	4.30	30	19.53	德国	70	4.53	12	5.46
10	英国	117	3.05	20	11.00	西班牙	31	2.00	9	8.32

天然橡胶主产国中除印尼和越南外，泰国、马来西亚、印度和中国SCI论文发文量逐年递增，1996—2000年，印度、马来西亚和泰国天然橡胶研究SCI论文发文量远远超过中国，2001—2010年，中国天然橡胶SCI论文发文量大幅增长，进入前10名，仅次于印度、马来西亚和泰国，到2011—2015年时，中国天然橡胶SCI论文发文量跃居世界第1位（图3-2）。这说明中国天然橡胶研究的SCI文献数量有了大幅提升，天然橡胶及其相关学科研究有了长足的发展。

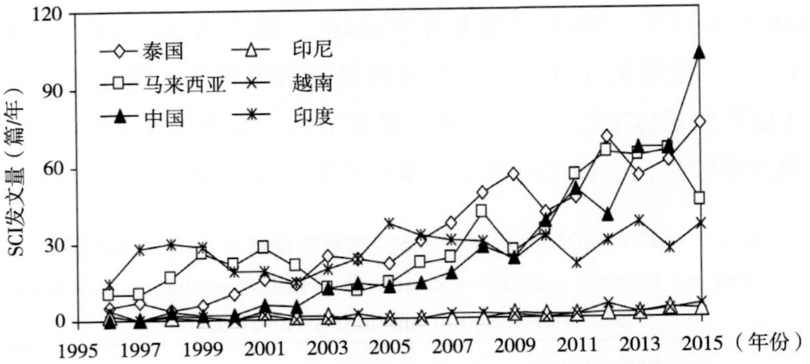

图3-2　1996—2015年主要产胶国天然橡胶SCI收录文献数量变化

Fig. 3-2　The change of the number of papers of nature rubber major countries published in SCI-indexed journals during 1996-2015

3.5　国际天然橡胶论文的主要发文机构

1996—2015年，天然橡胶研究被SCI收录文献涉及1 786个研究机构，排名前10位的科研机构主要分布在马来西亚、泰国、中国、印度和法国，这与发文量排名靠前的国家一致。天然橡胶SCI论文发文量排名靠前的研究机构有马来西亚理科大学（排名第1，303篇），泰国的宋卡王子大学（排名第2，229篇）、玛希隆大学（排名第3，215篇）和朱拉隆功大学（排名第7，149篇），中国热带农业科学院（排名第4，161篇），海南大学（排名第17，65篇），印度的甘地大学、印度橡胶研究所、法国农业发展研究中心和法国国家科学研究中心发文量都达100篇以上（表3-2）。这说明世界天然橡胶研究SCI论文的发文机构以大学为主，科研院所次之，中国天然橡胶研究

SCI论文发文机构以中国热带农业科学院和海南大学为代表。

表3-2 1996—2015年天然橡胶SCI收录文献数排名前10位的研究机构

Tab. 3-2 The top 10 most productive institutions of nature rubber papers published in SCI-indexed journals during 1996-2015

排序 Rank	研究机构 Research institution	记录数 Record count	百分比（%） Percent（%）
1	马来西亚理科大学（University Sains Malaysia）	303	7.89
2	宋卡王子大学（Prince Songkla University）	229	5.96
3	玛希隆大学（Mahidol University）	215	5.60
4	中国热带农业科学院（Chinese Academy of Tropical Agricultural Sciences）	161	4.19
5	甘地大学（Mahatma Gandhi University）	148	3.85
6	法国农业发展研究中心（CIRAD）	140	3.65
7	朱拉隆功大学（Chulalongkorn University）	139	3.62
8	法国国家科学研究中心[Centre National de la Recherche Scientifique（CNRS）]	112	2.92
9	印度橡胶研究所（Rubber Research Institute of India）	100	2.60
10	印度理工学院[Indian Institute of Technology（IIT）]	89	2.32

3.6 国际天然橡胶论文的主要学科分布

1996—2015年，天然橡胶研究被SCI收录文献涉及116个学科，排名前10名的学科中以高分子科学发文量最多，达2 123篇，占论文总数的55.29%，发文量达200篇以上的有学科交叉材料科学、化学工程、植物科学和合成材料等学科，发文量达100篇以上的有物理化学、生物化学与分子生物学、过敏学、材料科学特性测试和化学学科等（表3-3）。中国天然橡胶研究的学科主要有高分子科学、植物科学和材料学等。说明世界天然橡胶研究以加工领域的高分子科学、材料学和化学为主，农业领域的植物科学、生物化学和分子生物学为辅，还涉及医学领域的过敏学，中国天然橡胶研究学科与世界基本同步。

表3-3 1996—2015年天然橡胶SCI收录文献数排名前10位的学科分布

Tab. 3-3 The top 10 most productive subjects of nature rubber papers published in SCI-indexed journals during 1996-2015

排序 Rank	学科 Subject	记录数 Record count	百分比（%） Percent（%）
1	高分子科学（Polymer Science）	2 123	55.29
2	材料科学及交叉科学（Materials Science Multidiscipinary）	370	9.64
3	化学工程（Engineering Chemical）	282	7.34
4	植物科学（Plant Sciences）	231	6.02
5	合成材料（Materials Science Composites）	220	5.73
6	物理化学（Chemistry Physical）	174	4.53
7	生物化学与分子生物学（Biochemistry Molecular Biology）	167	4.35
8	过敏学（Allergy）	120	3.13
9	材料科学表征与测试（Materials Science Characterization Testing）	119	3.10
10	化学及交叉科学（Chemistry Multidisciplinary）	106	2.76

3.7 国际天然橡胶论文的主要来源期刊以及影响因子分布

1996—2015年，天然橡胶研究被SCI收录文献涉及688个期刊，发文量最多的期刊主要集中在美国和马来西亚，美国的Journal of Applied Polymer Science发文量最多（546篇），其次是Rubber Chemistry and Technology（145篇），马来西亚的Journal of Rubber Research（107篇）排名第3，发文量排名前10的期刊主要分布在高分子科学、材料科学和工程研究领域，期刊影响因子最高为3.562，最低为0.212（表3-4）。说明天然橡胶研究的SCI论文主要发表刊物的影响因子较低，发达国家和较早开展天然橡胶研究的主产国国家的发表刊物较多。

表3-4 1996—2015年天然橡胶SCI收录文献数排名前10位的出版物

Tab. 3-4 The top 10 most productive journals of nature rubber papers published in SCI-indexed journals during 1996-2015

排序 Rank	出版物 Journal name	记录数 Record count	百分比（%） Percent（%）	影响因子 Impact factor	研究领域 research field	国家 Country
1	应用高分子科学杂志（Journal of Applied Polymer Science）	546	14.22	1.768	高分子科学	美国
2	橡胶化学与技术（Rubber Chemistry and Technology）	145	3.78	1.024	高分子科学	美国
3	橡胶研究杂志（Journal of Rubber Research）	107	2.79	0.225	天然橡胶综合研究	马来西亚
4	高分子塑料技术与工程（Polymer Plastics Technology and Engineering）	102	2.66	1.481	高分子科学	美国
5	聚合物（Polymer）	90	2.34	3.562	高分子科学	英国
6	聚合物试验（Polymer Testing）	88	2.29	2.24	材料科学/高分子科学	英国
7	欧洲聚合物杂志（European Polymer Journal）	68	1.77	3.005	高分子科学	英国
8	弹性体与塑料杂志（Journal of Elastomers and Plastics）	64	1.67	0.773	高分子科学/材料科学	英国
9	橡胶塑料（KGK-Kautschuk Gummi Kunststoffe）	60	1.56	0.212	高分子科学/工程	德国
10	高分子工程与科学（Polymer Engineering and Science）	49	1.28	1.520	高分子科学/工程	美国

3.8 国际天然橡胶论文的主要基金资助机构

1996—2015年，天然橡胶研究被SCI收录文献共有1 600个基金资助机构，2 369篇文献不含基金资助，基金资助率为38.32%。排名前10的基金资助机构主要分布在中国、泰国、马来西亚和巴西，中国国家自然科学基金资助论文数最多（175篇），其次是泰国基金研究会（122篇），马来西亚理科大学（52篇）排名第3，巴西国家科学技术发展委员会、圣保罗研究基金会、中国的现代农业产业技术体系建设专项资金资助论文比例也较多（表3-5）。这说明天然橡胶研究SCI论文基金资助率较低，主要产胶国对天然橡胶研究SCI论文的资助力度大。

表3-5　1996—2015年天然橡胶SCI收录文献数排名前10位的基金资助机构

Tab. 3-5　The top 10 most productive foundations of nature rubber papers published in SCI-indexed journals during 1996-2015

排序 Rank	基金资助机构 Foundation name	记录数 Record count	百分比（%） Percent（%）
1	中国国家自然科学基金（National Natural Science Foundation of China）	175	4.55
2	泰国研究基金（Thailand Research Fund）	122	3.18
3	马来西亚理科大学（University Sains Malaysia）	52	1.35
4	巴西国家科学技术发展委员会（CNPQ）	33	0.86
5	巴西圣保罗研究基金会（FAPESP）	33	0.86
6	现代农业产业技术体系建设专项资金（Earmarked Fund for Modern Agro-Industry Technology Research System）	32	0.83
7	宋卡王子大学（Prince of Songkla University）	29	0.76
8	巴西高等教育人员促进会（CAPES）	25	0.65
9	日本科学促进学会（Japan Society for the Promotion of Science）	19	0.50
10	马来西亚橡胶董事会（Malaysian Rubber Board）	18	0.47

3.9　进入ESI前1%学科的天然橡胶高被引论文

Essential Science Indicators（基本科学指标，简称ESI）是一个基于Web of Science数据库的深度分析研究工具。ESI可以确定在某个研究领域有影响力的国家、机构、论文和出版物，以及研究前沿（吴明智和高硕，2015）。1996—2015年，入选ESI前1%学科的天然橡胶研究的高被引论文有5篇，主要分布在工程类、材料科学、农业科学、植物与动物科学研究领域；涉及的研究机构主要有印度的安那大学、马来西亚理科大学、法国科学研究中心、海南大学和中国热带农业科学院等；发文期刊主要有Journal of Hazardous Materials（危险材料）、Composites Science and Technology（复合材料科学与技术）、Chemical Engineering Journal（化学工程学报）、Industrial Crops and Products（经济作物和产品）以及Plant Molecular Biology（植物分子生物学），5种期刊分布在JCR分区的Q1区（表3-6）。这说明天然橡胶SCI论文在ESI收录的22个研究领域的影响力还较小，但发表刊物质量较高，中国天然橡胶研究的SCI论文已经进入世界ESI前1%之列。

表3-6 1996—2015年天然橡胶进入ESI前1%学科的高被引论文

Tab. 3-6 Ranked at top 1% subject in ESI database in nature rubber highly cited papers during 1996-2015

文献标题 Article titles	学科 Subject	被引频次 Citation rates	出版物 Journal name	影响因子 Impact factor	研究机构 Research institution
Chromium（VI）adsorption from aqueous solution by *Hevea Brasilinesis* sawdust activated carbon	Engineering	254	Journal of Hazardous Materials	4.529	Anna University
Short natural-fibre reinforced polyethylene and natural rubber composites：Effect of silane coupling agents and fibres loading	Materials Science	215	Composites Science and Technology	3.569	LMSE，Faculté des sciences de Sfax
Adsorption studies of basic dye on activated carbon derived from agricultural waste：*Hevea brasiliensis* seed coat	Engineering	115	Chemical Engineering Journal	4.321	Universiti Sains Malaysia
Mechanical，barrier，and biodegradability properties of bagasse cellulose whiskers reinforced natural rubber nanocomposites	Agricultural Sciences	91	Industrial Crops and Products	2.837	Centre National de la Recherche Scientifique（CNRS）
RNA-Seq analysis and de novo transcriptome assembly of *Hevea brasiliensis*	Plant & Animal Science	63	Plant Molecular Biology	4.257	Hainan University/ Chinese Academy of Tropical Agricultural Sciences

3.10 本章小结

（1）国际天然橡胶研究的发展动态。从时间序列、国家/地区两个层面对SCI-E数据库收录天然橡胶领域论文的整体情况进行对比分析发现，天然橡胶研究的发展动态特点如下：从时间序列上看，1996—2015年天然橡胶研究被SCI收录文献共计3 840篇，被引频次总计42 819次，发文量及引文量基本呈现逐年增加的趋势（图3-1）。总体而言，1996—2015年天然橡胶研究的发展动态良好，其研究实力和被关注程度在不断提升，研究成果也得到了较高的引用。但是高被引论文的数量仍然很少（表3-6）。从文献分布的国家/地区看，1996—2015年泰国、马来西亚、印度和中国等主要产胶国SCI论文数量排名世界前4位。德国、美国、法国和日本等发达国家文献的篇均他引次数排在前4位（表3-1，图3-2）。由此可见，天然橡胶的研究主要集中在天然橡胶主产国，然而发达国家仍然具备较高的研究水平。与其他国家比较，近5年来中国天然橡胶SCI论文发文量跃居世界第1位（图3-2），中国国家自然科学基金在所有资助机构中资助的论文最多，并逐渐形成了主要以中国热带农业科学院和海南大学为代表的研究机构。

（2）国际天然橡胶领域的学科分布。1996—2015年，天然橡胶领域的主要研究热点分布在高分子科学、材料科学、化学工程、植物科学和合成材料五大学科。另外，物理化学、生物化学与分子生物学、过敏学、材料科学特性测试和化学学科等学科也是天然橡胶领域重要的研究热点（表3-3）。中国天然橡胶研究的热点主要分布在高分子科学、植物科学、材料学等学科，中国天然橡胶研究热点与世界基本同步。橡胶树是热区典型的经济作物，在热带农业中天然橡胶具有特殊和重要的地位，主要分布在东南亚的发展中国家。由于这些国家的科研实力较弱，虽然天然橡胶研究的发文量在不断增加，但是仍然没有形成导向性的世界研究热点。世界天然橡胶研究热点依然集中在欧、美等发达国家，在橡胶采后加工及改性领域继续引导全球天然橡胶研究。中国天然橡胶研究起步较晚，高影响力论文偏少，但研究实力在不断增加，研究成果也逐渐被国际认可。中国应当继续增加资金投入，提高天然橡胶以及相关学科的科研水平。

（3）基于SCI-E数据库分析的总体概况。1996—2015年，世界天然橡胶研究发文量及引文量呈现逐年增加的趋势，发文量以天然橡胶主产国为主，而发达国家论文的篇均他引次数最高；主要发文机构分布在马来西亚、泰国、中国、印度和法国等国家；世界天然橡胶研究主要分布在高分子科学、材料学和化学、植物科学、生物化学和分子生物学等学科；发文量最多的期刊主要集中在美国和马来西亚；中国、泰国、马来西亚和巴西的论文基金资助数量位居前列；天然橡胶研究领域仅有5篇论文进入世界ESI前1%的高被引论文。

参考文献

刘彬，邓秀新，2015. 基于文献计量的园艺学基础研究发展状况分析[J]. 中国农业科学，48（17）：3 504–3 514.

孙秀焕，路文如，2012. 基于Web of Science的水稻研究态势分析[J]. 中国水稻科学，26（5）：607–614.

吴明智，高硕，2015. 基于ESI的中国农业科学十年发展态势的文献计量分析[J]. 农业图书情报学刊，27（9）：69–72.

中国热带农业科学院，2014. 中国热带作物产业可持续发展研究[M]. 北京：科学出版社. 71–76.

Bartol T，Mackiewicz-Talarczyk M，2015. Bibliometric analysis of publishing trends in fiber crops in Google Scholar，Scopus，and Web of Science[J]. Journal of Natural Fibers，12（6）：531–541.

Bornmann L，Mutz R，2015. Growth rates of modern science：a bibliometric analysis based on the number of publications and cited references[J]. Journal of the Association for Information Science and Technology，66（11）：2 215–2 222.

Cañas-Guerrero I，Mazarrón F R，Pou-Merina A，et al.，2013. Bibliometric analysis of research activity in the "Agronomy" category from the Web of Science，1997–2011[J]. European Journal of Agronomy，50：19–28.

De Winter J C F，Zadpoor A A，Dodou D，2014. The expansion of Google Scholar versus Web of Science：a longitudinal study[J]. Scientometrics，98（2）：1 547–1 565.

Finardi U，2013. Correlation between journal impact factor and citation performance：an experimental study[J]. Journal of Informetrics，7（2）：357–370.

Goodwin C，2014. Web of Science[J]. The Charleston Advisor，16（2）：55–61.

Hirsch J E，Buela-Casal G，2014. The meaning of the h-index[J]. International Journal of Clinical and Health Psychology，14（2）：161.

Ho Y S，Kahn M，2014. A bibliometric study of highly cited reviews in the Science Citation Index Expanded™[J]. Journal of the Association for Information Science and Technology，65（2）：372–385.

Knudson D，2014. Citation rates for highly-cited papers from different sub-disciplinary areas within kinesiology[J]. Chronicle of Kinesiology in Higher Education，25（2）：9–17.

Miyairi N，Chang H W，2012. Bibliometric characteristics of highly cited papers from Taiwan，2000–2009[J]. Scientometrics，92（1）：197–205.

Von Mark V C，Dierig D A，2012. Trends in literature on new oilseed crops and related species：seeking evidence of increasing or waning interest[J]. Industrial Crops and Products，37（1）：141–148.

Waltman L，Van Eck N J，2012. The inconsistency of the h-index[J]. Journal of the American Society for Information Science and Technology，63（2）：406–415.

4　国内天然橡胶文献计量分析

4.1　引言

中国知网，是中国国家知识基础设施（China National Knowledge Infrastructure，简称CNKI）的概念，于1998年由世界银行提出，于1999年由清华大学和清华同方发起建立。CNKI采用自主开发并具有国际领先水平的数字图书馆技术，建成了世界上全文信息量规模最大的“CNKI数字图书馆”及CNKI网络资源共享平台。该平台是涵盖期刊、博硕、年鉴、工具书、报纸、会议、标准、科技成果、专利及其他特色资源和国外资源的学科总库，其信息内容经过深度加工、编辑、整合，并以数据库形式进行有序管理，内容有明确的来源和出处，是学术研究与科学决策的重要依据（杨梓等，2014；马捷等，2011；董颖等，2014；郭金子，2014；王光和王利华，2015；张鹤凡和杨之音，2015；韩小莉等，2009）。文献计量学（Bibliometrics）是用数学和统计学的方法对文献进行定量分析的一个交叉学科，是由英国情报学家普里查德（Pritchard）于1969年提出（Pritchard，1969；吴爱芝，2016）。文献计量学既可以用于科研人员、高等院校、科研机构、期刊、国家、地区等的学术水平、科研能力和影响力等的评价，也可以用于某个研究领域的学科基础、发展态势、研究前沿和研究热点等的分析（刘亚伟和葛敬民，2013；仉晓红，2015）。例如，吴士蓉和王宁分别基于CNKI分析转基因玉米、超级水稻研究论文的研究现状与发展趋势；刘敏娟等利用SCI和CNKI数据库，采用文献计量学方法分析我国作物学核心期刊论文的研究态势；王彦等基于CSCD和CNKI等数据库对盐生植物文献进行统计分析（吴士蓉，2014；王宁，2014；刘敏娟等，2015；王彦和田长彦，

2013）。橡胶是热带地区典型的经济作物，我国科研人员围绕橡胶开展了大量的研究工作（中国热带农业科学院，2015；韩丹棠，1994），然而以天然橡胶为代表的主要热带作物文献计量研究相对滞后。本章采用文献计量学的方法对1996—2015年CNKI收录的天然橡胶科技文献进行统计分析，综述天然橡胶研究的发展状况，为热带农业科研工作者及相关决策者提供科学参考。

4.2 数据来源与研究方法

4.2.1 数据来源

本章的文献数据来源于CNKI的中国学术期刊网络出版总库（China Academic Journal Network Publishing Database，CAJD），该数据库收录国内学术期刊8 181种，全文文献总量46 479 214篇。以学术、技术、政策指导、高等科普及教育类期刊为主，内容覆盖自然科学、工程技术、农业、哲学、医学、人文社会科学等各个领域，是世界上最大的连续动态更新的中国学术期刊全文数据库（马兴，2015；王丽华，2014；谭菁，2014）。

4.2.2 研究方法

为保证数据获取的查准率，选择CNKI期刊数据库（CAJD）的高级检索功能，年限设置为1996—2015年，检索时间是2016年4月，检索式为篇名=（橡胶树+巴西橡胶+天然橡胶+天然胶乳+植胶区+橡胶园+橡胶林+割胶）*关键词=（橡胶树+巴西橡胶+天然橡胶+天然胶乳+植胶区+橡胶园+橡胶林+割胶），进行精确检索，并选择中英文扩检，共得到检索记录3 148条。将全部检索记录导入Excel，逐条查看，去除与本研究相关性较差的简讯、摘译、通知、声明、会议摘要以及重复文献，得到相关文献2 538篇。对这些相关文献从发文量、被引频次、发文机构、作者、来源期刊、影响因子、基金资助机构、关键词和高被引论文等方面进行统计分析。

4.3 国内天然橡胶论文的总体概况

1996—2015年CNKI收录天然橡胶文献总计2 538篇，包括1 126篇核心期刊论文，40篇SCI来源期刊论文，114篇EI来源期刊论文，30篇CSSCI期刊论文，外文文献53篇，中文文献2 485篇。其中，2 000篇论文被引用，占文献总数的78.80%，被引频次总计15 151次，每篇文献平均引用次数5.97次，单篇被引次数最高为119次。1996—2015年，年发文量呈现逐年增加趋势，被引频次也逐年增加（2013年后被引频次下降，是由于新发表论文的被引频次少）。1996—2007年，天然橡胶研究CNKI论文发文量较少，共796篇，占论文总数量的31.36%，被引频次7 695次；自2008年开始，天然橡胶研究的CNKI论文发文量大幅提升，至2015年总计1 742篇，占论文总数量的68.64%，被引频次7 456次（图4-1）。这说明了天然橡胶研究领域的关注程度、研究实力以及参考价值在不断提升。

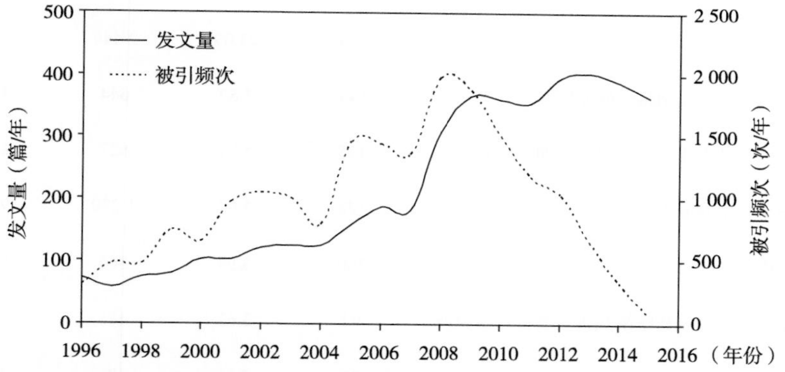

图4-1　1996—2015年天然橡胶CNKI论文发文量和被引频次

Fig. 4-1　The number and citation amount of nature rubber papers published in CNKI journals during 1996–2015

4.4　国内天然橡胶论文的主要发文机构

1996—2015年CNKI收录的天然橡胶文献中，排名前10位的科研机构发文量累计2 184篇，占期刊发表论文总数的86.05%；发文量和总被引频次最高的是中国热带农业科学院橡胶研究所；篇均被引频次最高的是中国科学院西双版纳热带植物园；另外，海南大学、云南热带作物科学研究所、中国热带农业科学院农产品加工研究所、华南热带农业大学、华南理工大学、热带生物技术研究所、青岛科技大学、中国热带农业科学院环境与植物保护研究所的发文量、总被引频次和篇均被引频次也进入前10名（表4-1）。这一研究领域的发文机构主要分布在三大植胶区（海南、云南和广东），并形成以中国热带农业科学院橡胶研究所和海南大学为领头的发文机构。

表4-1　1996—2015年天然橡胶CNKI收录文献数排名前10位的发文机构

Tab. 4-1　The top 10 most productive institutions of nature rubber papers published in CNKI journals during 1996–2015

排序 Rank	发文机构 Research institution	发文量 Total publications	百分比（%）Percent（%）	总被引频次 Sum of times cited	篇均被引频次 Average cited times citations
1	中国热带农业科学院橡胶研究所	794	31.28	3 867	4.87
2	海南大学	586	23.09	2 444	4.17
3	云南省热带作物科学研究所	149	5.87	644	4.32
4	中国热带农业科学院农产品加工研究所	145	5.71	627	4.32
5	华南热带农业大学	128	5.04	1 230	9.61
6	华南理工大学	100	3.94	864	8.64
7	中国热带农业科学院热带生物技术研究所	92	3.62	488	5.30
8	青岛科技大学	79	3.11	381	4.82
9	中国热带农业科学院环境与植物保护研究所	68	2.68	392	5.76
10	中国科学院西双版纳热带植物园	43	1.69	806	18.74

4.5　国内天然橡胶论文的主要发文作者和第一作者

如表4-2所示，1996—2015年CNKI收录天然橡胶文献中，发文量排前10名的作者包括第一作者、通讯作者和合作者发表的论文。这些作者的发文量累计694篇，占论文总数的27.34%；主要发文作者相对集中，所属机构包括中国热带农业科学院橡胶研究所、华南理工大学和海南大学，其中，发文量和总被引频次最高的是中国热带农业科学院橡胶研究所的黄华孙研究员，其发表的论文涉及林业、农艺学和生物学等学科，篇均被引频次最高的是华南理工大学的贾德民教授，论文学科是有机化工、化学和材料科学。

表4-2 1996—2015年天然橡胶CNKI收录文献数排名前10位的作者

Tab. 4-2 The top 10 most authors of nature rubber papers published in CNKI journals during 1996-2015

排序 Rank	作者 Authors	所属机构 Organizations	发文量 Total publications	总被引频次 Sum of times cited	篇均被引频次 Average cited times citations
1	黄华孙	中国热带农业科学院橡胶研究所	123	679	5.52
2	林位夫	中国热带农业科学院橡胶研究所	95	554	5.83
3	谢贵水	中国热带农业科学院橡胶研究所	86	566	6.58
4	李维国	中国热带农业科学院橡胶研究所	67	310	4.63
5	校现周	中国热带农业科学院橡胶研究所	66	372	5.64
6	贾德民	华南理工大学	63	564	8.95
7	廖双泉	海南大学	53	233	4.40
8	田维敏	中国热带农业科学院橡胶研究所	49	207	4.22
9	华玉伟	中国热带农业科学院橡胶研究所	47	184	3.91
10	杨礼富	中国热带农业科学院橡胶研究所	45	281	6.24

发文量排前10名的第一作者发文量累计134篇，占该研究领域期刊发表论文总数的5.28%。这10位作者所属机构均为中国热带农业科学院，其中，发文量最多的作者是吴春太，总被引频次最高的作者是吴志祥，篇均被引频次最高的作者是林位夫（表4-3）。

表4-3　1996—2015年天然橡胶CNKI收录文献数排名前10位的第一作者

Tab. 4-3　The top 10 most first author of nature rubber papers published in CNKI journals during 1996–2015

排序 Rank	第一作者 First author	所属机构 Institution	发文量 Total publications	总被引频次 Sum of times cited	篇均被引频次 Average cited times citations
1	吴春太	中国热带农业科学院橡胶研究所	20	65	3.25
2	吴志祥	中国热带农业科学院橡胶研究所	16	177	11.06
3	邹　智	中国热带农业科学院橡胶研究所	15	85	5.67
4	叶德林	云南天然橡胶产业股份有限公司	14	12	0.86
5	曹建华	中国热带农业科学院橡胶研究所	13	126	9.69
6	林位夫	中国热带农业科学院橡胶研究所	12	151	12.58
7	祁栋灵	中国热带农业科学院橡胶研究所	12	56	4.67
8	曾宗强	中国热带农业科学院农产品加工研究所	11	72	6.55
9	李德军	中国热带农业科学院橡胶研究所	11	16	1.45
10	校现周	中国热带农业科学院橡胶研究所	10	116	11.60

4.6　国内天然橡胶论文的主要来源期刊

1996—2015年CNKI收录的天然橡胶文献中涉及396种期刊，包含12种外文期刊和384种中文期刊，排名前10位期刊的发文量总计1 308篇，占期刊文献总数的51.54%。发文量较多的期刊主要集中在海南、云南、北京、甘肃、陕西、吉林；研究领域涉及农业综合、农业经济、有机化工和工业经济。发文量和总被引频次最高的是《热带作物学报》，篇均被引频次最高的是《合成橡胶工业》，综合影响因子最高的是《中国农学通报》（表4-4）。《热带作物学报》是由中国热带作物学会主办，中国热带农业科学院承办的综合性学术期刊，《合成橡胶工业》是由中国石油兰州石化公司主办的高分子弹性体材料科学与工程技术领域的专业性学术刊物，目前两种期刊均被中国科学引文数据库（CSCD）与北京大学《中文核心期刊要目总览》数据库收录。

表4-4 1996—2015年天然橡胶CNKI收录文献数排名前10位的发文期刊

Tab. 4-4 The top 10 most Publications of nature rubber papers published in CNKI journals during 1996-2015

排序 Rank	来源期刊 Source journal	发文量 Total publications	百分比（%） Percent（%）	综合影响因子 Impact factor	总被引频次 Sum of times cited	篇均被引频次 Average cited times citations
1	热带作物学报	372	14.66	0.509	2 617	7.03
2	热带农业科学	233	9.18	0.362	1 257	5.39
3	热带农业科技	155	6.11	0.085	564	3.64
4	弹性体	97	3.82	0.337	554	5.71
5	合成橡胶工业	97	3.82	0.349	868	8.95
6	特种橡胶制品	80	3.15	0.258	540	6.75
7	中国农学通报	77	3.03	0.566	455	5.91
8	橡胶工业	72	2.84	0.261	206	2.86
9	中国热带农业	65	2.56	无	246	3.78
10	热带农业工程	60	2.36	0.151	210	3.50

4.7 国内天然橡胶论文的基金资助情况

1996—2015年CNKI收录的天然橡胶文献中，2 478篇文献含基金资助，60篇文献不含基金资助，基金资助率为97.64%。排名前10的基金资助机构发文量总计1 208篇，占期刊文献总数的47.60%。以国家自然科学基金资助论文数最多（579篇），其次是海南省自然科学基金（317篇），国家科技支撑计划（94篇）排名第3位，国家重点基础研究发展计划（973计划）、云南省自然科学基金、国家高技术研究发展计划（863计划）、社会公益研究专项计划、广东省自然科学基金资助论文比例也较多（表4-5）。说明天然橡胶研究论文的基金资助率高，国家级和省级基金对天然橡胶研究论文的资助力度大。

表4-5　1996—2015年天然橡胶CNKI收录文献数排名前10位的基金资助机构

Tab. 4-5　The top 10 most funding agency of nature rubber papers published in CNKI journals during 1996-2015

排序 Rank	基金资助机构 Funding agency	发文量 Total publications	百分比（%） Percent（%）
1	国家自然科学基金	579	22.81
2	海南省自然科学基金	317	12.49
3	国家科技支撑计划	94	3.70
4	国家重点基础研究发展计划（973计划）	49	1.93
5	云南省自然科学基金	37	1.46
6	国家高技术研究发展计划（863计划）	34	1.34
7	社会公益研究专项计划	32	1.26
8	广东省自然科学基金	30	1.18
9	海南省教育厅科研基金	20	0.79
10	国家社会科学基金	16	0.63

4.8　国内天然橡胶论文中的高被引论文

1996—2015年CNKI收录的天然橡胶文献中，中国热带农业科学院橡胶研究所有4篇论文进入天然橡胶研究领域被引频次排名前10位，中国科学院西双版纳热带植物园有3篇，青岛科技大学、北京化工大学和中国热带农业科学院热带生物技术研究所各1篇（表4-6）；发文期刊主要有《植物生态学报》《生态学报》《植物生理学报》《热带作物学报》《热带农业科学》《合成橡胶工业》《复合材料学报》《遗传》，综合影响因子最高为1.813，最低为0.349；被引频次最高的是中国科学院西双版纳热带植物园的房秋兰于2006年发表在《植物生态学报》上的论文，被引频次达119次。

表4-6 1996—2015年天然橡胶CNKI收录文献数排名前10位的高被引论文

Tab. 4-6 The top 10 most highly cited of nature rubber papers published in CNKI journals during 1996-2015

排序 Rank	文献标题 Title	所属机构 Institution	被引频次 Times cited	来源期刊 Source journal	综合影响因子 Impact factor
1	西双版纳热带季节雨林与橡胶林土壤呼吸	中国科学院西双版纳热带植物园	119	植物生态学报	1.813
2	西双版纳季节雨林与橡胶多层林凋落物动态的比较研究	中国科学院西双版纳热带植物园	116	植物生态学报	1.813
3	西双版纳地区热带季节雨林与橡胶林林冠水文效应比较研究	中国科学院西双版纳热带植物园	86	生态学报	1.604
4	一种提取橡胶树叶中总DNA的方法	中国热带农业科学院橡胶研究所	70	植物生理学报	0.826
5	茉莉酸刺激的橡胶树胶乳cDNA消减文库的构建及其序列分析	中国热带农业科学院橡胶研究所	65	热带作物学报	0.509
6	海南垦区胶园肥力演变探研	中国热带农业科学院橡胶研究所	60	热带农业科学	0.362
7	核磁共振法表征硫黄用量对天然橡胶交联密度及结构的影响	青岛科技大学	59	合成橡胶工业	0.349
8	橡胶树EST-SSR标记的开发与应用	中国热带农业科学院橡胶研究所	58	遗传	0.902
9	离体培养橡胶树体细胞诱导纯多倍性无性系方法的研究初报	中国热带农业科学院热带生物技术研究所	53	热带作物学报	0.509
10	碳纳米管/天然橡胶复合材料的制备及性能	北京化工大学	48	复合材料学报	0.835

4.9 国内天然橡胶论文的高频关键词分析

通过关键词分析可以发现天然橡胶的热点研究领域。1996—2015年CNKI收录天然橡胶文献中，排除检索中使用的关键词，天然橡胶文献中出现频次200次以上的高频关键词有硫化胶、拉伸强度、拉断伸长率、硫化特性、硫化体系、干胶、力学性能、定伸应力、橡胶种植、排胶；出现100次以上的高频关键词有胶乳制品、胶联密度、橡胶分子、橡胶复合材料、物理性能、混炼胶、正硫化、寒害、天然橡胶产业、乙烯利、乳管细胞、橡胶加工（表4-7）。说明天然橡胶研究热点主要集中在3个领域：一是天然橡胶加工、有机高分子材料领域；二是橡胶树产胶生理、生物化学与分子生物学领域；三是天然橡胶产业经济领域。

表4-7 1996—2015年天然橡胶CNKI收录文献的高频关键词

Tab. 4-7 The top 10 most high frequency keywords of nature rubber papers published in CNKI journals during 1996-2015

词频 Occurrences	关键词 Keyword	词频 Occurrences	关键词 Keyword
492	硫化胶	163	橡胶分子
455	拉伸强度/拉断伸长率	136	橡胶复合材料
311	硫化特性/硫化体系	136	物理性能
296	力学性能	131	混炼胶
292	干胶含量/干胶	125	正硫化时间
262	定伸应力	121	寒害
254	橡胶种植	120	天然橡胶产业
220	排胶	115	乙烯利
170	胶乳制品	107	乳管细胞
166	交联密度	101	橡胶加工

4.10 本章小结

1996—2015年，CNKI收录天然橡胶文献总计2 538篇，年发文量呈现逐年增加的趋势。主要发文机构集中分布在三大植胶区，并形成以中国热带农业科学院橡胶研究所和海南大学为领头的发文机构。发文期刊大多是以热带作物和橡胶为特色的专业性学术刊物；天然橡胶研究论文受国家自然科学基金和海南省自然科学基金的资助率最高；天然橡胶研究热点主要分布在加工、有机高分子材料、产胶生理、生物化学与分子生物学和产业经济等领域。

参考文献

董颖，于海霞，张艳芳，等，2014. 基于学生视角的CNKI数据库检索平台评价[J]. 情报科学，32（8）：85-90.

郭金子，2014. 基于CNKI数据库的文献计量分析工具研究[J]. 图书馆学刊，36（4）：113-116，122.

韩丹棠，1994. 我国天然橡胶研究文献的统计分析[J]. 农业图书情报学刊（3）：21-25.

韩小莉，李恩科，康延兴，等，2009. Google学术搜索及其与CNKI检索功能的对比[J]. 情报杂志，28（z2）：182-183，199.

仉晓红，2015. 文献计量方法应用进展浅析[J]. 河北科技图苑，28（4）：44-47.

刘敏娟，王婷，袁雪，等，2015. 基于文献计量学的中国机构作物学科竞争力分析[J]. 农业展望，11（3）：59-65.

刘亚伟，葛敬民，2013. 发表于图情核心期刊的文献检索课研究论文的计量分析[J]. 情报科学，31（4）：115-118，160.

马捷，刘小乐，郑若星，2011. 中国知网知识组织模式研究[J]. 情报科学，29（6）：843-846.

马兴，2015. 中文资源的收集、整理与利用初探——以CAJD数据库为例[J]. 办公自动化（综合版）（23）：38-39，24.

谭菁，2014. 基于CNKI的华中农业大学期刊论文的统计分析[J]. 农业图书情报学刊，26（1）：60-68.

王光，王利华，2015. 知网、万方和维普数据库检索平台对农业科技文献检索结果分析[J]. 农业网络信息（7）：59-61.

王丽华，2014. CNKI改版的启示——以CAJD为例[J]. 农业网络信息（10）：94-97.

王宁，2014. 基于文献的我国转基因大豆研究发展态势分析[J]. 大豆科学，33（5）：764-767.

王彦，田长彦，2013. 基于CSCD的盐生植物研究文献计量评价[J]. 植物分类与资源学报，35（5）：665-673.

吴爱芝，2016. 信息技术进步与文献计量学发展[J]. 现代情报，36（2）：32-37.

吴士蓉，2014. 中国转基因玉米研究论文文献计量学分析[J]. 农业图书情报学刊，26（4）：50-53.

杨梓，刘冰，吴菲菲，等，2014. 用户体验的创新研究——以中国知网新旧版功能比较为例[J]. 情报杂志，33（3）：202-207.

张鹤凡，杨之音，2015. 农业领域文献增长规律及发展趋势研究[J]. 科技情报开发与经济，25（17）：122-124.

中国热带农业科学院，2014. 中国热带作物学科发展研究[M]. 北京，科学出版社. 155-164.

Pritchard A，1969. Statistical bibliography or bibliometrics?[J]. Journal of Documentation，25（4）：348-349.

5 国际天然橡胶研究热点分析

5.1 引言

文献计量学是国际公认的图书情报领域重要分支学科，已发展到对文献内部知识单元进行计量研究阶段（Régibeau and Rockett，2016；Corrall et al.，2013；邱均平等，2008；苏新宁，2013）。科学知识图谱是文献计量领域的创新研究方法，以可视化图形展示科学知识内部的动态演进过程，揭示其研究前沿、研究热点和结构关系等（Chen，2012；Chen and Leydesdorff，2014；李杰和陈超美，2016）。CiteSpace是目前应用最广泛的信息可视化工具，它利用文献引文网络的知识可视化分析原理，在一幅图谱上呈现一个知识域的知识基础、演化历程、发展趋势和最新动向（陈悦等，2015；胡泽文等，2013；赵丹群，2012）。Liu等对1985—2014年全球水稻研究文献进行计量分析，基于关键词共词网络揭示水稻领域的研究热点，还绘制了园艺植物高被引论文文献共被引图谱；吴同亮等分析2016年国内外环境土壤学的研究热点；Liu等对作物生长模型SCI文献进行主题共现、国家/地区合作网络、作者和文献共被引分析；Chen等分析了WoS收录的农业科技创新领域关键节点文献的研究前沿和演进路径（Liu et al.，2017；刘彬和邓秀新，2015；吴同亮等，2017；Liu et al.，2014；Chen et al.，2016）。

橡胶是热区典型的经济作物，天然橡胶是国家重要的战略物资，天然橡胶在热带农业中具有重要地位（国家天然橡胶产业技术体系，2016；中国热带农业科学院，2015）。但热带农业文献计量分析相对薄弱，因此以天然橡胶为代表的科学知识图谱计量分析亟待开展。本章利用文献计量方法结合CiteSpace技术，对2002—2016年Web of Science收录的天然橡胶研究主题文献进行关键词聚类分析，以可视化的方式展示国际天然橡胶领域的研究热点，期望为天然橡胶研究跟踪国际前沿、把握研究方向及规划科研项目提供一定的科学量化依据。

5.2 数据来源与研究方法

5.2.1 数据来源

数据来源于Web of Science核心合集的Science Citation Index Expanded（SCI-E）数据库。本章定义的天然橡胶特指从巴西三叶橡胶树提取的天然橡胶，并包括银胶菊（*Parthenium argentatum*）、蒲公英（*Taraxacum brevicorniculatum*）等天然胶乳；本章的天然橡胶文献包括天然橡胶种植及其相关的分子生物学研究方法，不涉及天然橡胶加工研究。因此，采用表5-1制定的检索策略进行主题检索，时间跨度为2002—2016年，文献类型为“Article”，得到文献记录1 724条（检索日期为2017年2月6日）。按照CiteSpace软件要求导出文本格式数据，并采用数据过滤与除重功能进行预处理。

表5-1 天然橡胶研究主题文献检索策略

Tab. 5-1 Retrieval strategy of topical articles on natural rubber

检索式 Retrieval type	文献数 Records	检索策略 Topic search
#1	7 992	主题：（“rubber tree*” or “*Hevea*” or “natur* rubber” or “natur* latex” or “nr latex” or “rubber latex”）
#2	3 106	主题：（“rubber”）and主题：（“tapping” or “plantation*” or “yard” or “garden” or “forest” or “NR” or “planting area*” or “planting” or “growing area*” or “growing state”）
#3	1 724	#1or #2 精炼依据：Web of Science类别（Entomology or Plant Sciences or Soil Science or Biochemical Research Methods or Biochemistry Molecular Biology or Environmental Sciences or Forestry or Agronomy or Biotechnology Applied Microbiology or Microbiology or Agriculture Multidisciplinary or Agricultural Engineering or Ecology or Genetics Heredity）即生物学、植物科学、土壤科学、生物化学研究方法、生物化学与分子生物学、环境科学、林业、农学、生物技术与应用微生物、微生物学、农业及相关科学、农业工程、生态学、遗传学

5.2.2 研究方法

采用CiteSpace（5.0.R3.SE版本）进行关键词共现聚类分析。CiteSpace相关参数设置，时区分割（Timing Slicing）：分别设置为2002—2006、2007—2011、2010—2016；时间切片（Years Per Slice）：5；节点类型（Node Types）：Keyword；节点阈值（Selection Criteria）：Top 100；网络修剪方式（Pruning）：Pathfinder（寻径算法）和Pruning sliced networks（修剪每个切片网络），其余选项均为默认。分别得到3个时期的天然橡胶研究共词网络图谱。图谱中每个节点代表一个关键词，节点的大小表示关键词出现的次数，节点之间连线的粗细表示关键词共现强度的高低，彼此邻近的关键词表示它们通常出现在相似的文献中（Chen，2012；Chen and Leydesdorff，2014；李杰和陈超美，2016；陈悦等，2015）。图中节点年轮的颜色和厚度，表示关键词出现的时间（本次分析将5年分割为一个时间段，因此只出现蓝色）和数量。具有紫色外环的节点，具有高中介中心性，外环越厚表示该节点位于关键词网络较为中心的位置，是网络中各部分连接的过渡。

5.3 2002—2006年天然橡胶研究热点分析

按照软件相应设置，得到节点数量107个，连线数量128条，关键词形成4个网络聚类的可视化图谱。该时期研究热点可分为“橡胶树天然橡胶生物合成与调控”“橡胶树产排胶及防御关键蛋白与基因”“橡胶林生物多样性、生态效应与土壤微生物碳”“天然橡胶胶乳蛋白过敏”（图5-1）。

5.3.1 橡胶树天然橡胶生物合成与调控

基因表达（gene expression）、克隆（clone）、植物（plant）等为该聚类中的关键节点，与这些节点相连的关键词有橡胶生物合成（rubber biosynthesis）、聚异戊二烯（polyisoprene）、转移酶活性（transferase activity）、延伸因子（elongation factor）、粒子（particle）、分子量（molecular weight）等，表明参与天然橡胶生物合成转移酶类和橡胶粒子是研究热点。Asawatreratanakul等从橡胶树胶乳中分离得到2个*CPT*基因*HRT1*和*HRT2*，发现*HRT2*在体外催化合成的天然橡胶分子量已经接近植物体内合成的天然橡胶分子量大小，推测*HRT2*可能是橡胶转移酶（Asawatreratanakul et al.，2003）。Takaya等从橡胶树叶片和胶乳cDNA文库中克隆到*GGPS*基因，分别命名为*HbGGPPS1*和*HbGGPPS2*，该编码蛋白都含有反式异戊烯链延长酶特有的保守区（Takaya et al.，2003）。Singh等研究表明，抗小橡胶粒子蛋白（SRPP）抗体与*F.carica*和*F.benghalensis*的橡胶粒子均不存在免疫识别反应，认为这两种植物橡胶粒子的表面可能没有与SRPP同源的蛋白，而橡胶树的SRPP大小远远小于这两种植物（Singh et al.，2003）。聚类图中还出现镁（magnesium）、异戊烯焦磷酸盐（isopentenyl pyrophosphate）、银胶菊（*Parthenium argentatum*）等，说明天然橡胶生物合成调控也是关注热点。

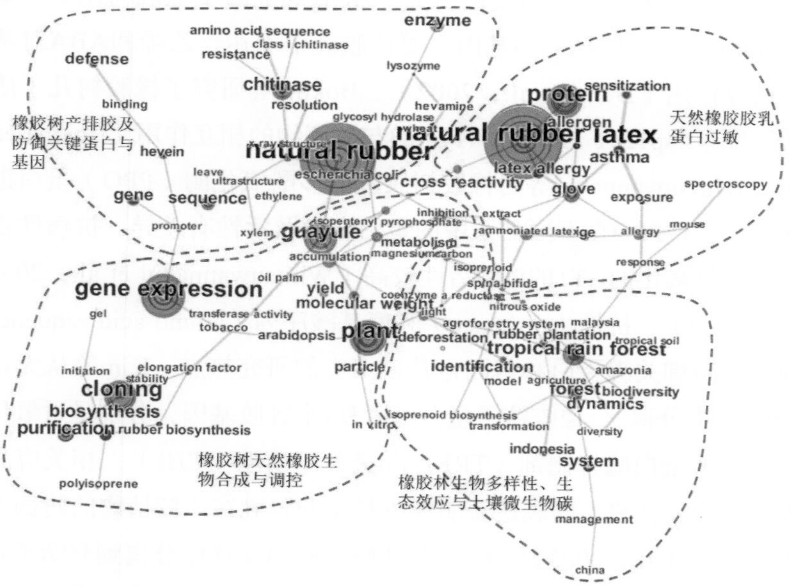

图5-1　2002—2006年天然橡胶研究主题文献关键词共现网络图谱

Fig. 5-1　Keyword co-occurring network map of topical articles on natural rubber during 2002-2006

（1）金属离子调节：Scott等研究发现，Mg^{2+}或Mn^{2+}浓度影响IPP掺入合成中的天然橡胶分子的速度；在橡胶树、银胶菊和印度榕3种植物中，橡胶转移酶能结合FPP、FPP-金属离子或单独结合金属离子，但不能单独结合IPP分子（Scott et al.，2003）。da Costa等发现在体外合成条件下，Mg^{2+}浓度可显著影响橡胶树和银胶菊所合成天然橡胶的分子量，而橡胶树对Mg^{2+}浓度更敏感（da Costa et al.，2005）。（2）温度调节：Cornish等发现夜晚约20℃的低温对成熟银胶菊植株橡胶转移酶活性有诱导作用（Cornish and Backhaus，2003）。（3）茉莉酸类物质调节：Tian等发现外施茉莉酸甲酯可以提高胶乳的干胶含量，表明茉莉酸类物质能够促进乳管中橡胶的生物合成（Tian et al.，2003）。

5.3.2　橡胶树产排胶及防御关键蛋白与基因

几丁质酶（chitinase）、橡胶几丁质酶（hevamine）、橡胶素（hevein）、基因（gene）、酶（enzyme）等关键词的出现，表明橡胶树还侧重于产排胶关键基因与蛋白研究。Deng等克隆了橡胶树*Hevein*基因，Northernblot分析表明*Hevein*基因主要在胶乳中表达，乙烯和ABA对基因的表达有诱导作用（Deng et al.，2002）。Bokma等研究了橡胶树几丁质酶催化区活性位点Asp125、Glu127和Tyr183与其底物的相互作用（Bokma et al.，2002）。Wititsuwannakul等对橡胶树胶乳的多酚氧化酶（PPO）蛋白进行分离和纯化，发现PPO在不同品系中的蛋白含量及活性有差异，抗病性强、胶乳及其产品易褐化的GT1的PPO活性较高（Wititsuwannakul et al.，2002）。防御（defence）、抗性（resitance）和氨基酸序列（amino acid sequence）等的出现说明防御关键蛋白和基因也是该聚类的研究热点。Kim等从无花果树（*F.carica*）中分离2个橡胶粒子蛋白基因和1个乳胶基因，分别是过氧化物酶（POX）、胰蛋白酶抑制剂（TRI）和i类几丁质酶（CHI），相关应激蛋白在橡胶粒子和胶乳表面的表达表明橡胶粒子和胶乳在产胶植物的防御中起重要作用（Kim et al.，2003）。Chen等从胶乳RNA文库中分离到*MYB*类转录因子，命名为*HbMyb1*；该转录因子在橡胶树正常叶片、树皮和胶乳中都表达，但在死皮橡胶树的树皮中表达量显著降低，说明*HbMyb1*是参与橡胶树死皮病发生的关键基因（Chen et al.，2003）。Sookmark等采用2-DE方法比较健康和死皮橡胶树胶乳蛋白质组的差异，发现相对分子质量为15kDa、22kDa的2种多肽在死皮树的细胞质中累积，经鉴定分别是橡胶延伸因子（REF）Hevb 1和小橡胶粒子蛋白（SRPP）Hevb 3（Sookmark et al.，2002）。

5.3.3 橡胶林生物多样性、生态效应与土壤微生物碳

该聚类中出现热带雨林（tropical rain forest）、橡胶林（rubber plantation）、农林生态系统（agroforestry system）、生物多样性（biodiversity）、多样性（diversity）等关键词，代表橡胶林生物多样性研究。Beukema等比较了苏门答腊低湿地原始森林、橡胶复合农林生态系统和橡胶林的蕨类植物在地块和土地空间水平的多样性。发现3种土地利用类型的地块平均物种丰富度差异不显著；橡胶农林系统的物种—面积曲线具有更高的边坡参数，表明橡胶农林系统的β多样性更高；并将陆生蕨类植物划分为“森林物种”和“非森林物种”，橡胶农林系统中等数量的“森林物种”比橡胶林多，但少于原始森林，因此种植更大面积的橡胶林有利于保护物种多样性（Beukema and van Noordwijk，2004）。Jones等从印度尼西亚苏门答腊岛低湿地7种土地利用类型（原始森林、砍伐森林、成熟橡胶农林系统、成熟橡胶林、南洋楹种植园、白茅草地、木薯种植园）收集白蚁种群栖息地的相关变量数据，研究白蚁群落多样性下降原因（Jones et al.，2003）。Martius等比较了亚马孙中部冠层封闭的原始林、12年次生林、农林系统、橡胶林、桃棕榈单作的小气候特征，发现发达的冠层有效保护了土壤大型底栖动物免受高温和干旱胁迫，优化这些冠层封闭的农林系统有助于更好地管理有益的土壤分解者群落（Martius et al.，2004）。一氧化氮（nitrous oxide）、类异戊二烯生物合成（isoprenoid biosynthesis）、森林砍伐（deforestation）、模型（model）和光（light）等的出现表明橡胶林气体排放、光合特性等生态效应也是研究热点。Ishizuka等测定了4种土地利用类型（老龄林、过度采伐林、采伐后烧荒以及橡胶林）CO_2、CH_4和N_2O的排放。发现森林管理和采伐、森林转变为人工林显著影响森林土壤温室气体排放（Ishizuka et al.，2002）。Baker等在旱季和雨季对云南西双版纳人工橡胶林和热带次生林冠层上空的异戊二烯和单萜烯排放通量进行连续测定。雨季测到异戊二烯排放通量的最大值，日平均值为1mg·Cm^{-2}·h^{-1}，旱季异戊二烯通量下降明显为0.15mg·Cm^{-2}·h^{-1}（Baker et al.，2005）。Senevirathna等比较自然遮阴下控制光线强度对橡胶树生长、光合特性和暗适应的影响，发现橡胶树通过暗适应和遮阴诱导的动态光抑制作用降低而加强在遮阴条

件下的早期生长（Senevirathna et al.，2003）。另外，碳（carbon）、热带土壤（tropical soil）和有机质（organic matter）等代表橡胶林土壤微生物碳研究热点。Kurzatkowski等研究了巴西亚马孙地区桃棕榈单作、橡胶林单作和含有4个树种的农林系统的凋落物分解、土壤微生物的生物量和活性（Kurzatkowski et al.，2004）。Zhang等分析了中国海南植胶区土壤微生物碳和总有机碳（Zhang and Zhang，2003）。

5.3.4 天然橡胶胶乳蛋白过敏

该聚类关键词主要有天然橡胶胶乳（natural rubber latex）、蛋白（protein）、手套（glove）、过敏（allergy）、提取（extract）和抑制（inhibition）等。Tomazic-Jezic等开发了一种ELISA抑制方法，用于提取和定量天然橡胶胶乳产品蛋白，该方法测定的抗原蛋白含量更接近于提取物中总蛋白质含量（Tomazic-Jezic et al.，2002）。Palosuo等基于捕获ELISA的单克隆抗体和纯化或重组变应原方法检测天然橡胶胶乳过敏原。结果表明，Hevb 6.02和Hevb 5是成人的2种主要过敏原，Hevb 3和Hevb 1是脊柱裂患儿的2种主要过敏原（Palosuo et al.，2002）。Charous等发现长期接触乳胶产品粉末或胶乳气溶胶的工作人员是职业性天然橡胶胶乳过敏患者，最常见的症状是接触性荨麻疹，吸入可能会导致过敏性鼻炎和哮喘；皮肤刺痛测试是诊断NRL过敏最准确方法（Charous et al.，2002）。Meade等采用小鼠模型评估天然橡胶胶乳蛋白过敏，以降低胶乳产品的过敏原和致敏个体对过敏原的交叉反应力（Meade and Woolhiser，2002）。

5.4 2007—2011年天然橡胶研究热点分析

该时期得到节点数量133个，连线数量151条，关键词形成5个网络聚类的图谱。研究热点分为“橡胶树杂交育种与分子辅助育种”“橡胶树胶乳再生及其对外界刺激的分子响应”“橡胶树乳管关键蛋白及其基因表达”“产胶植物天然橡胶产量与质量”“橡胶林土壤有机碳、内生菌多样性与病害”（图5-2）。

图5-2 2007—2011年天然橡胶研究主题文献关键词共现网络图谱

Fig. 5-2 Keyword co-occurring network map of topical articles on natural rubber during 2007–2011

5.4.1 橡胶树杂交育种与分子辅助育种

该聚类出现遗传多样性（genetic diversity）、多态性（polymorphism）和种质（germplasm）等关键词，反映研究人员对橡胶树种质杂交育种的关注程度。Arantes等以橡胶树30个混合授粉—自交和异交—无性系后代群体为材料，对其干胶产量、茎围长度的遗传增益进行评估。鉴定了20个最佳个体，分别获得67.96%和16.48%的遗传增益，近交系数约为2.82%（Arantes et al.，2010）。Gouvêa等研究了22个IAC400基因型橡胶树的遗传变异与分化，利用产量等 8 个数量性状进行单因素和多因素分析。并采用多因素分析和SSR标记研究60个IAC基因型橡胶树的遗传多样性（Gouvêa et al.，2010a；2010b）。Diaby等采用多目标方法选育橡胶树无性系，确定包括割胶、15年总产量、15～25年总产量、抗风性、抗病性、生理抗性、嫁接、质量等选育标准（Diaby et al.，2010）。体细胞胚胎发生（somatic embryogenesis）、培养（culture）和诱导（induction）等代表橡胶树遗传转化分子辅助育种研究热点。Hua等通过次生体细胞胚发生对橡胶树自根幼态无性系（CATAS 7-33-97和CATAS 88-13）进行快繁（Hua et al.，2010）。Leclercq等利用绿色荧光蛋白（GPF）作为橡胶树根癌脓杆菌介导转化的有效选择标记（Leclercq et al.，2010）。

5.4.2 橡胶树胶乳再生及其对外界刺激的分子响应

该聚类出现橡胶生物合成（rubber biosynthesis）、聚异戊二烯生物合成（isoprenoid biosynthesis）、橡胶粒子（rubber particle）和克隆（cloning）等关键词，说明天然橡胶合成、产排胶和胶乳再生仍是科学家持续关注的研究热点。Rojruthai等利用^{14}C标记研究胶乳中小橡胶粒子（SRP）在天然橡胶体外合成中的作用（Rojruthai et al.，2010）。Li等分析橡胶树硫氧还蛋白*h*基因（*HbTRX1*）在自根幼态无性系与其供体老态无性系胶乳中的差异表达。推测该基因可能影响胶乳流动和胶乳产量差异（Li et al.，2011）。Deng等克隆和表征了橡胶树肌动蛋白解聚因子（ADF）基因，表达谱分析表明*HbADF*可能与乳胶再生和流动有关（Deng et al.，2010）。而乙烯利（ethephon）、茉莉酸甲酯（jasmonic acid）、途径（pathway）和转录组分析（transcriptome analysis）等代表橡胶树胶乳再生对乙烯利、茉莉酸甲酯等外界刺激的分子响应。Duan等对外界伤害、茉莉酸甲酯和乙烯利刺激响应的一组信号和代谢途径中所涉及基因的转录丰度进行研究，揭示橡胶树响应外界刺激的25个基因的表达模式（Duan et al.，2010）。Tian等克隆和表征了橡胶树乳管细胞中与茉莉酸信号途径相关的*HbJAZ1*基因，初步证明乳管细胞橡胶生物合成主要受茉莉酸信号调控（Tian et al.，2010）。Zhu等克隆和分析了橡胶树钙调蛋白基因*HbCDPK1*功能。发现机械损伤、茉莉酸甲酯和乙烯利诱导*HbCDPK1*表达上调，推测该基因可能与胶乳产量和橡胶生物合成有关（Zhu et al.，2010）。

5.4.3 橡胶树乳管关键蛋白及其基因表达

该聚类出现乳管（laticifer）、纯化（purification）、天然橡胶胶乳（nature rubber latex）、蛋白（protein）和质膜（plasma membrane）等关键词。Tang等鉴定了一个乳管中高丰度表达的蔗糖转运蛋白基因*HbSUT3*的功能，发现乙烯利和割胶可以诱导该基因的表达，与胶乳产量正相关，由此推断*HbSUT3*是决定乳管蔗糖供给、影响橡胶产量的关键*SUT*基因（Tang et al.，2010）。Tungngoen等研究施用乙烯利、生长素、脱落酸和水杨酸对提高橡胶树胶乳产量和2个水通道蛋白基因表达差异的影响。结果表明，乙烯利处理可上调*HbPIP2; 1*在胶乳和树皮中的表达，上调*HbTIP1; 1*在胶乳表达但下调其在树皮中的表达，下调*HbPIP1; 1*在胶乳和树皮中的表达；*HbPIP2; 1*和*HbTIP1; 1*对其他3种激素的反应与乙烯利类似，*HbPIP1; 1*可轻微受生长素诱导，但不受脱落酸和水杨酸调控（Tungngoen et al.，2011）。Dusotoit-Coucaud等利用cDNA文库筛选方法克隆橡胶树多元醇转运蛋白基因*HbPLT1*和*HbPLT2*，发现*HbPLT1*和*HbPLT2*的表达模式响应于不同的刺激而变化；与*HbPLT1*相比，乙烯利刺激和机械伤口诱导乳管和内部树皮细胞中*HbPLT2*显著上调；并结合蔗糖转运蛋白（Tang et al.，2010）和水通道蛋白（Tungngoen et al.，2011）的表达和定位情况，提出该质粒膜蛋白参与刺激产胶（Dusotoit-Coucaud et al.，2010a；2010b）。

5.4.4 产胶植物天然橡胶产量与质量

该聚类出现银胶菊（*Parthenium argentatum*）、天然胶乳（nature rubber latex）、胶乳产量（latex yield）、质量（quality）和性能（performance）等关键词。Coffelt等进行银胶菊生长周期中胶乳产量和生物量的最佳收割时期试验（Coffelt et al.，2010）。Salvucci等为了明确限制产胶因素，在模拟夏季和冬季的温度和光周期下诱导银胶菊的营养活跃期和休眠期，对其在营养活跃期和休眠期的碳水化合物和类异戊二烯产物的光合作用和同化分配进行研究（Salvucci et al.，2010）。Pearson等评估了商业化种植向日葵的生物量及其生物量的分配，确定了胶乳浓度、产量、品种遗传多样性以及影响胶乳质量的因素。并开发了一种适用于低分子量产胶植物的胶乳提取方法（加速溶剂萃取），用于提取和定量向日葵天然胶乳（Pearson et al.，2010a；

2010b）。Buranov等采用混合法和流动法分别提取哈萨克斯坦和乌兹别克斯坦的橡胶草（蒲公英）胶乳，并利用溶剂循序提取法进行橡胶草固体胶的提取（Buranov et al.，2009）。

5.4.5 橡胶林土壤有机碳、内生菌多样性与病害

该聚类出现橡胶林（rubber plantation）、森林（forest）、热带雨林（tropical rain forest）、农林系统（agroforestry system）、土地利用（land use）、碳（carbon）、有机质（organic matter）等关键词，表明该时期的研究热点由橡胶林土壤微生物碳向有机碳转变。Saha等比较了印度喀拉拉邦的5种土地利用类型（庭院、自然林、单一种植的椰子林、水稻田和橡胶林）的土壤碳储量，发现有高密度树木的如森林和小尺寸的庭院的碳储量较高（Saha et al.，2010）。Moreira等研究了亚马孙中部地区不同植物覆盖（原始森林、次生林、牧场、柑橘林、橡胶林）下的土壤肥力、矿质氮和微生物量，发现橡胶林是土壤有机质变化最小的植物覆盖（Moreira et al.，2011）。Zhang等研究了入侵蚯蚓对云南西双版纳南部橡胶林土壤活性炭的影响，发现蚯蚓可以沿着土壤垂直剖面重新分布活性炭，保护土壤表层积累的LOC，而土壤下层的积累的LOC不受保护（Zhang et al.，2010）。多样性（diversity）、生物多样性（biodiversity）、菌类（fungi）和孢子（microcyclus ulei）等代表该时期还出现橡胶林内生菌多样性研究。Gazis等分离和鉴定了橡胶树叶片和茎段的内生真菌分离物，其主要分布于子囊菌属、青霉属、担子菌门、接合菌门等（Gazis et al.，2010）。Guyot等研究了从野生橡胶树分离获得橡胶南美叶疫病菌菌株的方法（Guyot et al.，2010）。Diniz等进行AMF菌根菌对橡胶树苗木生长、生物物理参数和解剖学方面影响的试验（Diniz et al.，2010）。而疾病（disease）、致病性（pathogenicity）、生物防治（biological control）和抗性（resistance）等表明橡胶树病害也是该聚类研究热点之一，主要涉及对南美叶疫病、棒孢霉落叶病的抗性和防治药剂筛选等（Rivano et al.，2010；Moraes et al.，2011；Fernando et al.，2010a，2010b）。树皮（bark）、毁林（deforestationde）还代表了橡胶树死皮病研究。de Faÿ等分析死皮病橡胶树氰化作用的发生及其对次生韧皮部生理和病理的影响，并比较健康和死皮病橡胶树树皮韧皮部的

超微结构特征（De Faÿ et al.，2010；2011）。Pramod等比较死皮病橡胶树树皮和健康树皮形成层结构的变化，发现患病树皮梭形细胞经历变形分裂，乙烯利刺激加速健康树皮梭形细胞的细胞分裂（Pramod et al.，2011）。

5.5 2012—2016年天然橡胶研究热点分析

该时期可视化图谱节点与连线数量均为122，关键词形成5个网络聚类，分别是“产胶植物天然橡胶生物合成”“橡胶树分子标记与miRNAs”“橡胶树防御蛋白对非生物胁迫的分子响应”“热带雨林转变为橡胶林的土地利用变化效应”以及“天然橡胶胶乳微生物降解”（图5-3）。

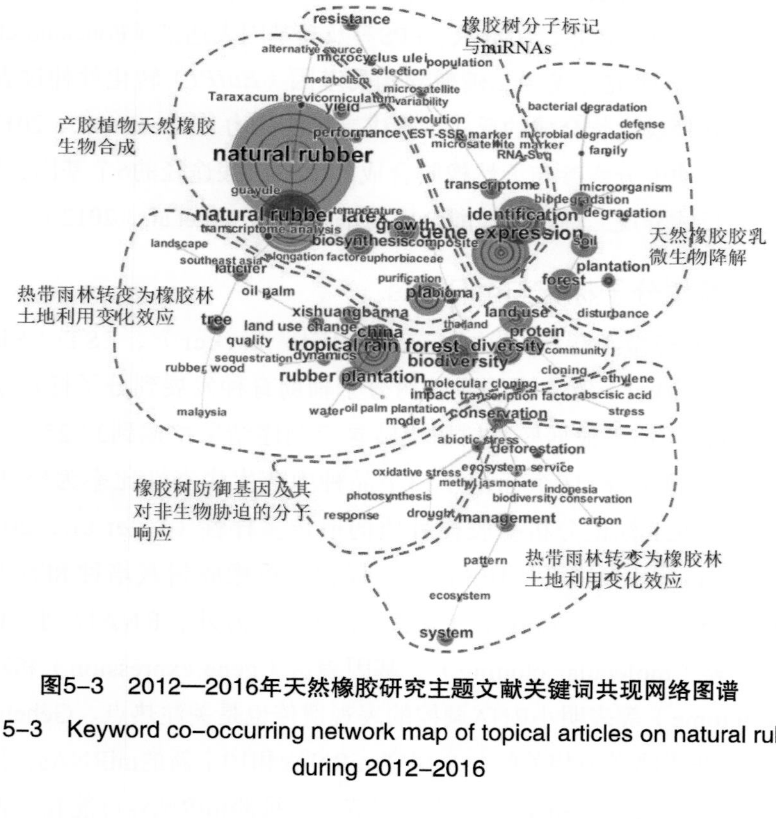

图5-3 2012—2016年天然橡胶研究主题文献关键词共现网络图谱

Fig. 5-3 Keyword co-occurring network map of topical articles on natural rubber during 2012-2016

5.5.1 产胶植物天然橡胶生物合成

该聚类出现银胶菊（*Parthenium argentatum*）、蒲公英（*Taraxacum*

brevicorniculatum）、替代来源（alternative source）、天然胶乳（natural rubber latex）、生物合成（biosynthesis）等关键词，表明该时期研究热点由橡胶树天然橡胶生物合成转向产胶植物天然橡胶生物合成。van Deenen等克隆了短角蒲公英甲羟戊酸途径的3–羟基–3–甲基–戊二酸酰辅酶A合酶（HMGS）、3–羟基–3–甲基戊二酰辅酶A还原酶（HMGR）（van Deenen et al.，2012）。Collins-Silva等通过抑制俄罗斯蒲公英中*TkSRPP3*的表达，发现橡胶含量和聚合长度均明显下降。证明SRPP在橡胶粒子结构稳定和天然橡胶合成中均起重要作用（Collins-Silva et al.，2012）。Post等研究短角蒲公英乳管中的*CPT*，显著抑制乳管橡胶粒子形成和橡胶生物合成（Post et al.，2012）。Ponciano等研究低温诱导下银胶菊中天然橡胶生物合成途径的转移酶活性，如HMGS、HMGR、FPS等及其基因表达谱（Ponciano et al.，2012）。Chen等研究异戊烯二磷酸异构酶基因（*EuIPI*）转化杜仲过表达植株，产生天然橡胶生物合成的反式聚异戊二烯的能力（Chen et al.，2012）。Suzuki等从杜仲中分离参与天然橡胶合成的甲羟戊酸途径的6个基因，以及5个编码全长反式异戊二烯基二磷酸合成酶基因（Suzuki et al.，2012）。

5.5.2 橡胶树分子标记与miRNAs

该聚类出现微卫星标记（microsatellite marker）、EST–SSR标记（EST-SSR marker）等，说明橡胶树分子辅助育种发展到分子标记遗传多样性研究阶段。Li等根据橡胶树树皮转录组测序结果挖掘到39 257个EST–SSR标记，随机选择110个标记在13个品种中产生多态性比率为55.45%，还运用EST–PCR标记分析橡胶树种质的遗传多样性（Li et al.，2012a；2012b）。Perseguini等利用EST–SSR标记分析橡胶树栽培种和野生型种质的遗传多样性（Perseguini et al.，2012）。另外，RNA序列（RNA-Seq）、克隆（molecular cloning）、基因表达（gene expression）和转录组（transcriptome）等表明小RNA调控的表观遗传也是关注热点。Gébelin等鉴定与橡胶树非生物逆境相关的48个保守miRNAs和10个新的miRNAs，预测的靶基因涉及刺激应答、抗氧化和转录调节等，推测miRNAs可能在调控氧化还原中起作用（Gébelin et al.，2012）。Lertpanyasampatha等鉴定与橡胶树产量相关的56个保守miRNAs和20个新的miRNAs，这些靶基因涉及逆境响应、代谢反应和信号转导等（Lertpanyasampatha et al., 2012）。

5.5.3 橡胶树防御基因及其对非生物胁迫的分子响应

该聚类包含非生物胁迫（abiotic stress）、乙烯（ethylene）、茉莉酸甲酯（methyl jasmonate）、脱落酸（abscisic acid）、干旱（drought）、响应（response）等关键词。Zhang等分析了*HbWRKY1*基因的功能，发现外界损伤、乙烯利或茉莉酸刺激可以诱导胶乳中*HbWRKY1*基因表达，该基因与胁迫应答反应有关（Zhang et al., 2012）。Yang等从橡胶树中分离出编码14-3-3蛋白的cDNA，命名为*Hb14-3-3c*。通过酵母双杂交筛选，在橡胶树胶乳中鉴定了*Hb14-3-3c*的11个相互作用蛋白，认为参与了橡胶生物合成、应激、防御反应等（Yang et al., 2012）。Chen等克隆和鉴定了调节乙烯利和茉莉酸信号途径中发挥重要作用的AP2/ERF转录因子，命名为*HbEREBP1*，发现割胶、机械损伤、外源乙烯利和茉莉酸甲酯刺激诱导其表达量下调，推测*HbEREBP1*可能是防御基因家族成员的负调节因子（Chen et al., 2012）。

5.5.4 热带雨林转变为橡胶林土地利用变化效应

热带雨林（tropical rain forest）、土地利用变化（land use change）、橡胶林（rubber plantation）、生物多样性（biodiversity）、动力学（dynamics）和模型（model）等关键词，说明该时期橡胶林由生物多样性、生态效应与土壤有机碳转向热带雨林转变为橡胶林的土地利用变化效应研究。Fox等利用土地利用转化及其动力学效应模型（CLUE-s）对东南亚山地陆地的土地利用/覆被变化进行模拟研究。目前约有9%的覆被由当地的树木、灌木和草本植物组成，预计将在50年期间被树木、茶叶等常绿灌木取代，4%由扩张种植的橡胶树取代（Fox et al., 2012）。Boithias等在泰国东北部一个限水区，基于水分养分和光捕获农林系统（WaNuLCAS）动力学模型预测橡胶林水分利用、生长和胶乳产量（Boithias et al., 2012）。Chaudhuri等研究了印度热带雨林转变为橡胶林后，5种类型蚯蚓对土壤有机碳、磷钾利用率等的影响（Chaudhuri et al., 2012）。Moreira等评估了亚马孙中部酸性土壤和贫瘠土壤中土地利用类型转变下种植的包括橡胶在内的12种作物（木材、果树和棕榈）的土壤肥力和养分需求（Moreira et al.,

2012）。中国（China）出现频次较高，说明对中国热带雨林转变为橡胶林的土地利用变化效应是国际研究热点。Meng等调查了热带土地利用变化下云南南部橡胶林甲壳动物群落和物种分布，发现5年和8年的橡胶林没有显著指标值的物种分布，20年和40年的橡胶林物种种类减少。他们还调查了云南南部土地利用类型（天然林地、开放土地、农田和不同树龄橡胶林）变化下食蚜蝇和野生蜜蜂物种丰富度和群落组成结构（Meng et al.，2012a；2012b）。Li等评估了云南西双版纳大面积热带森林转变为橡胶林和茶叶种植园的土地利用变化下，土壤无机氮和坡度的变化（Li et al.，2012）。

5.5.5 天然橡胶胶乳微生物降解

该聚类包含生物降解（biodegradation）、细菌降解（bacterial degradation）、微生物降解（microbial degradation）、土壤（soil）和鉴定（identification）等关键词。Watcharaku等从土壤中分离鉴定出一株天蓝色链霉菌（*Streptomyces coelicolor*）菌株CH13，能够生物降解木薯淀粉/天然橡胶高分子聚合物和橡胶手套（Watcharakul et al.，2012）。Hesham等报道了嗜热芽孢杆菌菌株ASU7作为降解天然橡胶胶乳的有效来源，将编码16S rRNA的基因序列和系统发育分析应用于细菌属和种的鉴定（Hesham et al.，2012）。Shah等分离的新菌株AF-666被鉴定为芽孢杆菌属，是在含聚异戊二烯培养基上生长能力最强的菌株，可以降解聚异戊二烯橡胶（Shah et al.，2012）。Hiessl等通过基因组测序预测放线菌转化菌株VH2的天然胶乳生物降解途径（Hiessl et al.，2012）。

5.6 2002—2016年天然橡胶研究历程分析

2002—2016年共检索到天然橡胶研究主题文献1 724篇。从文献计量网络图谱（图5-3）和高频关键词列表（表5-2）可以看出，3个时期的节点数逐渐增多，关键词词频上升，表明天然橡胶研究的逐渐扩展和深入。2002—2016年，关键词中出现*Arabidopsis*、protein、plant、cloning、gene、identification、biosynthesis，并且guayule、biosynthesis关键词词频在2007—2016年上升，表明橡胶树天然橡胶生物合成与调控一直是研究热点，并转向产胶植物天然橡胶生物合成及产量、质量研究。2002—2006年，chitinase、 enzyme关键词的出现，表明早期以橡胶树防御蛋白和基因研究为主，后期逐渐转向橡胶树产排胶、胶乳再生和流动、非生物胁迫应答的关键蛋白和基因研究。tropical rain forest、rubber plantation、land use关键词的出现及词频的上升，反映出橡胶林生态效应、生物多样性和土壤有机碳也一直是科学家关注的研究热点，并逐渐转向热带雨林转变为橡胶林土地利用变化效应研究。另外，China、Xishuangbanna出现频次较高，说明对中国西双版纳橡胶林土地利用变化效应研究也是国际热点。

表5-2　2002—2016年天然橡胶研究主题文献前15位高频关键词

Tab. 5-2　Top 15 high frequency keywords of topical articles on natural rubber during 2002-2016

序号 Rank	2002—2006年 During 2002-2006	2007—2011年 During 2007-2011	2012—2016年 During 2012-2016
1	protein（27）	plant（47）	*Arabidopsis*（77）
2	plant（26）	gene expression（39）	gene expression（69）
3	cloning（26）	*Arabidopsis*（34）	tropical rain forest（62）
4	guayule（22）	growth（33）	plant（54）
5	tropical rain forest（20）	cloning（31）	land use（49）
6	growth（16）	identification（26）	identification（46）
7	chitinase（16）	protein（25）	growth（45）
8	purification（14）	guayule（21）	diversity（43）
9	somatic embryogenesis（13）	biosynthesis（21）	China（38）
10	enzyme（12）	rubber plantation（19）	biodiversity（38）
11	crystal structure（12）	tropical rain forest（18）	rubber plantation（35）
12	biosynthesis（12）	laticifer（18）	biosynthesis（33）
13	asthma（11）	diversity（15）	protein（29）
14	gene（10）	land use（15）	Xishuangbanna（28）
15	sequence（10）	crystal structure（15）	gene（28）

注：（ ）中数字表示高频关键词出现频次

Note：The number in（ ）represents the occurrence of high frequency keywords

5.7 本章小结

（1）从微观层面上，2002—2016年“天然橡胶生物合成与调控”主题始终是天然橡胶领域关注的研究热点之一。2002—2006年，“天然橡胶生物合成与调控”主题重点关注橡胶树天然橡胶生物合成关键蛋白和基因，以及金属、温度、茉莉酸类物质的辅助调节作用。乳管是天然橡胶合成的主要场所，2007—2011年，该主题重点关注橡胶树乳管关键蛋白及其基因表达；由于橡胶树受种植面积限制，难以满足全球日益增长的对天然橡胶的需求，因此该时期还涉及产胶植物天然橡胶质量和产量研究。随着美国等国家大规模扩大种植银胶菊、蒲公英（橡胶草）等产胶植物（国家天然橡胶产业技术体系，2016），2012—2016年，该主题重点关注产胶植物天然橡胶生物合成关键基因研究。另外，“橡胶树胶乳再生关键蛋白和基因”主题也是该领域持续关注的热点。该主题早期以橡胶树产排胶及防御关键蛋白与基因研究为主，后期逐渐转向橡胶树胶乳再生及其对外界刺激或非生物胁迫的分子响应。

（2）从宏观层面上，2002—2016年“橡胶林土壤碳、生物多样性和生态效应”也一直是该领域的研究热点。该主题由橡胶林土壤有机碳、生态效应与生物多样性研究，转向热带雨林转变为橡胶林的土地利用变化效应研究，并侧重于对中国西双版纳橡胶林土地利用变化效应研究。

（3）随着天然橡胶研究的不断扩大和深入，该研究领域还体现了多学科的交叉和融合。橡胶树杂交育种、遗传转化和分子标记等辅助育种技术日趋成熟。天然橡胶制品胶乳蛋白提取和检测是早期的研究热点，天然橡胶微生物降解主题是近年来新兴发展的热点领域。

参考文献

陈悦，陈超美，刘则渊，等，2015. CiteSpace知识图谱的方法论功能[J]. 科学学研究，33（2）：242-253.

国家天然橡胶产业技术体系，2016. 中国现代农业产业可持续发展战略研究天然橡胶分册[M]. 北京：中国农业出版社. 11-35.

胡泽文，孙建军，武夷山，2013. 国内知识图谱应用研究综述[J]. 图书情报工作，57（3）：131-137.

李杰，陈超美，2016. CiteSpace：科技文本挖掘及可视化[M]. 北京：首都经济贸易大学出版社. 194–203.

刘彬，邓秀新，2015. 基于文献计量的园艺学基础研究发展状况分析[J]. 中国农业科学，48（17）：3 504–3 514.

邱均平，杨瑞仙，陶雯，等，2008. 从文献计量学到网络计量学[J]. 评价与管理，4（2）：1–9.

苏新宁，2013. 文献计量学与科学评价中有关问题思考[J]. 图书与情报，149（1）：79–83.

吴同亮，王玉军，陈怀满，等，2017. 基于文献计量学分析2016年环境土壤学研究热点[J]. 农业环境科学学报，36（2）：205–215.

赵丹群，2012. 试论科学知识图谱的文献计量学研究范式[J]. 图书情报工作，56（6）：107–110.

中国热带农业科学院，2015. 中国热带作物学科发展研究[M]. 北京：科学出版社. 155-164.

Arantes F C，Gonçalves P S，Scaloppi Junior E J，et al.，2010. Genetic gain based on effective population size of rubber tree progenies[J]. Pesquisa Agropecuária Brasileira，45（12）：1 419–1 424.

Asawatreratanakul K，Zhang Y W，Wititsuwannakul D，et al.，2003. Molecular cloning，expression and characterization of cDNA encoding *cis*-prenyltransferases from *Hevea brasiliensis*[J]. European Journal of Biochemistry，270（23）：4 671–4 680.

Baker B，Bai J H，Johnson C，et al.，2005. Wet and dry season ecosystem level fluxes of isoprene and monoterpenes from a southeast Asian secondary forest and rubber tree plantation[J]. Atmospheric Environment，39（2）：381–390.

Beukema H，van Noordwijk M，2004. Terrestrial pteridophytes as indicators of a forest-like environment in rubber production systems in the lowlands of Jambi，Sumatra[J]. Agriculture，Ecosystems and Environment，104（1）：63–73.

Boithias L，Do F C，Ayutthaya S I N，et al.，2012. Transpiration，growth and latex production of a *Hevea brasiliensis* stand facing drought in Northeast Thailand：the use of the WaNuLCAS model as an exploratory tool[J]. Experimental Agriculture，48（1）：49–63.

Bokma E，Rozeboom H J，Sibbald M，et al.，2002. Expression and characterization of active site mutants of hevamine，a chitinase from the rubber tree *Hevea brasiliensis*[J]. European Journal of Biochemistry，269（3）：893–901.

Buranov A U，Elmuradov B J，2009. Extraction and characterization of latex and natural rubber from rubber-bearing plants[J]. Journal of Agricultural and Food Chemistry，58（2）：734–743.

Charous B L，Tarlo S M，Charous M A，et al.，2002. Natural rubber latex allergy in the occupational setting[J]. Methods，27（1）：15–21.

Chaudhuri P S，Pal T K，Nath S，et al.，2012. Effects of five earthworm species on some physico-chemical properties of soil[J]. Journal of Environmental Biology，33（4）：713.

Chen C M，2012. Predictive effects of structural variation on citation counts[J]. Journal of the American Society for Information Science and Technology，63（3）：431-449.

Chen C M，Leydesdorff L，2014. Patterns of connections and movements in dual-map overlays：a new method of publication portfolio analysis[J]. Journal of the Association for Information Science and Technology，65（2）：334–351.

Chen Q Q，Zhang J B，Huo Y，2016. A study on research hot-spots and frontiers of agricultural science and technology innovation–visualization analysis based on the CiteSpace Ⅲ[J]. Agricultural Economics-Zemedelska Ekonomika，62（9）：429–445.

Chen R，Harada Y，Bamba T，et al.，2012. Overexpression of an isopentenyl diphosphate isomerase gene to enhance trans-polyisoprene production in *Eucommia ulmoides* Oliver[J]. BMC Biotechnology，12（1）：78.

Chen S，Peng S，Huang G，et al.，2003. Association of decreased expression of a *Myb* transcription factor with the TPD（tapping panel dryness）syndrome in *Hevea brasiliensis*[J]. Plant Molecular Biology，51（1）：51–58.

Chen Y Y，Wang L F，Dai L J，et al.，2012. Characterization of *HbEREBP1*，a wound-responsive transcription factor gene in laticifers of *Hevea brasiliensis* Muell. Arg[J]. Molecular Biology Reports，39（4）：3 713–3 719.

Coffelt T A，Nakayama F S，2010. Determining optimum harvest time for guayule latex and biomass[J]. Industrial Crops and Products，31（1）：131–133.

Collins-Silva J，Nural A T，Skaggs A，et al.，2012. Altered levels of the *Taraxacum kok-saghyz*（Russian dandelion）small rubber particle protein，TkSRPP3，result in qualitative and quantitative changes in rubber metabolism[J]. Phytochemistry，79：46–56.

Corrall S，Kennan M A，Afzal W，2013. Bibliometrics and research data management services：emerging trends in library support for research[J]. Library Trends，61（3）：636–674.

Cornish K，Backhaus R A，2003. Induction of rubber transferase activity in guayule（*Parthenium argentatum* Gray）by low temperatures[J]. Industrial Crops and Products，17（2）：83–92.

da Costa B M T，Keasling J D，Cornish K，2005. Regulation of rubber biosynthetic rate and molecular weight in *Hevea brasiliensis* by metal cofactor[J]. Biomacromolecules，6（1）：279–289.

de Faÿ E，2011. Histo-and cytopathology of trunk phloem necrosis，a form of rubber tree（*Hevea brasiliensis* Müll. Arg.）tapping panel dryness[J]. Australian Journal of Botany，59（6）：563–574.

de Faÿ E，Moraes L A C，Moraes V H F，2010. Cyanogenesis and the onset of tapping panel dryness in rubber tree[J]. Pesquisa Agropecuária Brasileira，45（12）：1 372–1 380.

Deng X，Fei X，Huang J，et al.，2002. Isolation and analysis of rubber hevein gene and its promoter sequence[J]. Acta Botanica Sinica，44（8）：936–940.

Deng Z，Liu X，Chen C，et al.，2010. Molecular cloning and characterization of an actindepolymerizing factor gene in *Hevea brasiliensis*[J]. African Journal of Biotechnology，9（45）：7 603–7 610.

Diaby M，Valognes F，Clement-Demange A，2010. Amulticriteria decision approach for selecting *Hevea* clones in Africa[J]. Biotechnologie，Agronomie，Société et Environnement，14（2）：299–309.

Diniz P F A，Oliveira L E M，Gomes M P，et al.，2010. Growth，biophysical parameters and anatomical aspects of young rubber tree plants inoculated with arbuscular mycorrhizal fungi *Glomus clarum*[J]. Acta Botanica Brasilica，24（1）：65–72.

Duan C，Rio M，Leclercq J，et al.，2010. Gene expression pattern in response to wounding，methyl jasmonate and ethylene in the bark of *Hevea brasiliensis*[J]. Tree Physiology，30（10）：1 349–1 359.

Dusotoit-Coucaud A，Kongsawadworakul P，Maurousset L，et al.，2010a. Ethylene stimulation of latex yield depends on the expression of a sucrose transporter（*HbSUT1B*）in rubber tree（*Hevea brasiliensis*）[J]. Tree Physiology，30（12）：1 586–1 598.

Dusotoit-Coucaud A，Porcheron B，Brunel N，et al.，2010b. Cloning and characterization of a new polyol transporter（*HbPLT2*）in *Hevea brasiliensis*[J]. Plant and Cell Physiology，51（11）：1 878–1 888.

Fernando T，Jayasinghe C K，Wijesundera R L C，et al.，2010a. Evaluation of screening methods against Corynespora leaf fall disease of rubber（*Hevea brasiliensis*）[J]. Journal of Plant Diseases and Protection，117（1）：24–29.

Fernando T，Jayasinghe C K，Wijesundera R L C，et al.，2010b. Screening of fungicides against Corynespora leaf fall disease of rubber under nursery conditions/Screening von Fungizidwirkungen gegenüber der Corynespora-Blattfallkrankheit in Baumschulen[J]. Journal of Plant Diseases and Protection，117（3）：117–121.

Fox J，Vogler J B，Sen O L，et al.，2012. Simulating land-cover change in montane mainland southeast Asia[J]. Environmental Management，49（5）：968–979.

Gazis R, Chaverri P, 2010. Diversity of fungal endophytes in leaves and stems of wild rubber trees（*Hevea brasiliensis*）in Peru[J]. Fungal Ecology, 3（3）: 240–254.

Gébelin V, Argout X, Engchuan W, et al., 2012.Identification of novel microRNAs in *Hevea brasiliensis* and computational prediction of their targets[J]. BMC Plant Biology, 12（1）: 18.

Gouvêa L R L, Chiorato A F, Gonçalves P S, 2010a. Divergence and genetic variability among superior rubber tree genotypes[J]. Pesquisa Agropecuária Brasileira, 45（2）: 163–170.

Gouvêa L R L, Rubiano L B, Chioratto A F, et al., 2010b. Genetic divergence of rubber tree estimated by multivariate techniques and microsatellite markers[J]. Genetics and Molecular Biology, 33（2）: 308–318.

Guyot J, Doaré F, 2010. Obtaining isolates of *Microcyclus ulei*, a fungus pathogenic to rubber trees, from ascospores[J]. Journal of Plant Pathology, 92（3）: 765–768.

Hesham A E L, Mohamed N H, Ismail M A, et al., 2012. 16S rRNA gene sequences analysis of *Ficus elastica* rubber latex degrading thermophilic *Bacillus* strain ASU7 isolated from Egypt[J]. Biodegradation, 23（5）: 717–724.

Hiessl S, Schuldes J, Thürmer A, et al., 2012. Involvement of two latex-clearing proteins during rubber degradation and insights into the subsequent degradation pathway revealed by the genome sequence of *Gordonia polyisoprenivorans* strain VH2[J]. Applied and Environmental Microbiology, 78（8）: 2 874–2 887.

Hua Y W, Huang T D, Huang H S, 2010. Micropropagation of self-rooting juvenile clones by secondary somatic embryogenesis in *Hevea brasiliensis*[J]. Plant Breeding, 129（2）: 202–207.

Ishizuka S, Tsuruta H, Murdiyarso D, 2002. An intensive field study on CO_2, CH_4, and N_2O emissions from soils at four land-use types in Sumatra, Indonesia[J]. Global Biogeochemical Cycles, 16（3）: 1–11.

Jones D T, Susilo F X, Bignell D E, et al., 2003. Termite assemblage collapse along a land-use intensification gradient in lowland central Sumatra, Indonesia[J]. Journal of Applied Ecology, 40（2）: 380–391.

Kim J S, Kim Y O, Ryu H J, et al., 2003. Isolation of stress-related genes of rubber particles and latex in fig tree（*Ficus carica*）and their expressions by abiotic stress or plant hormone treatments[J]. Plant and Cell Physiology, 44（4）: 412–414.

Kurzatkowski D, Martius C, Höfer H, et al., 2004. Litter decomposition, microbial biomass and activity of soil organisms in three agroforestry sites in central Amazonia[J]. Nutrient Cycling in Agroecosystems, 69（3）: 257–267.

Leclercq J, Lardet L, Martin F, et al., 2010. The green fluorescent protein as an efficient selection marker for *Agrobacterium tumefaciens*-mediated transformation in *Hevea brasiliensis* Muell. Arg. [J]. Plant Cell Reports, 29（5）: 513–522.

Lertpanyasampatha M, Gao L, Kongsawadworakul P, et al., 2012. Genome-wide analysis of microRNAs in rubber tree（*Hevea brasiliensis* L.）using high-throughput sequencing[J]. Planta, 236（2）: 437–445.

Li D, Deng Z, Qin B, et al., 2012a. De novo assembly and characterization of bark transcriptome using Illumina sequencing and development of EST-SSR markers in rubber tree（*Hevea brasiliensis* Muell. Arg.）[J]. BMC Genomics, 13（1）: 192.

Li D, Xia Z, Deng Z, et al., 2012b. Development and characterization of intron-flanking EST-PCR markers in rubber tree（*Hevea brasiliensis* Muell. Arg.）[J]. Molecular Biotechnology, 51（2）: 148–159.

Li H, Ma Y, Liu W, et al., 2012. Soil changes induced by rubber and tea plantation establishment: comparison with tropical rain forest soil in Xishuangbanna, SW China[J]. Environmental Management, 50（5）: 837–848.

Li H L, Lu H Z, Guo D, et al., 2011. Molecular characterization of a thioredoxin *h* gene（*HbTRX1*）from *Hevea brasiliensis* showing differential expression in latex between self-rooting juvenile clones and donor clones[J]. Molecular Biology Reports, 38（3）: 1 989–1 994.

Liu B, Zhang L, Wang X, 2017. Scientometric profile of global rice research during 1985–2014[J]. Current Science, 112（5）: 1 003.

Liu H, Zhu Y, Guo Y, et al., 2014. Visualization analysis of subject, region, author, and citation on crop growth model by CiteSpace II software[M]. Knowledge Engineering and Management. Springer Berlin Heidelberg, 243–252.

Martius C, Höfer H, Garcia M V B, et al., 2004. Microclimate in agroforestry systems in central Amazonia: does canopy closure matter to soil organisms? [J]. Agroforestry Systems, 60（3）: 291–304.

Meade B J, Woolhiser M, 2002. Murine models for natural rubber latex allergy assessment[J]. Methods, 27（1）: 63–68.

Meng L Z, Martin K, Liu J X, et al., 2012a. Contrasting responses of hoverflies and wild bees to habitat structure and land use change in a tropical landscape（southern Yunnan, SW China）[J]. Insect Science, 19（6）: 666–676.

Meng L Z, Martin K, Weigel A, et al., 2012b. Impact of rubber plantation on carabid beetle communities and species distribution in a changing tropical landscape（southern Yunnan,

China）[J]. Journal of Insect Conservation，16（3）：423–432.

Moreira A，Fageria N K，Garciay Garcia A，2011. Soil fertility，mineral nitrogen，and microbial biomass in upland soils of the Central Amazon under different plant covers[J]. Communications in Soil Science and Plant Analysis，42（6）：694–705.

Moreira A，Moraes L A C，Fageria N K，2012. Nutritional limitations in multi-strata agroforestry system with native Amazonian plants[J]. Journal of Plant Nutrition，35（12）：1 791–1 805.

Moraes L A C，Moreira A，Fontes J R A，et al.，2011. Assessment of rubber tree panels under crowns resistant to South American leaf blight[J]. Pesquisa Agropecuária Brasileira，46（5）：466–473.

Palosuo T，Alenius H，Turjanmaa K，2002. Quantitation of latex allergens[J]. Methods，27（1）：52–58.

Pearson C H，Cornish K，McMahan C M，et al.，2010a. Natural rubber quantification in sunflower using an automated solvent extractor[J]. Industrial Crops and Products，31（3）：469–475.

Pearson C H，Cornish K，McMahan C M，et al.，2010b. Agronomic and natural rubber characteristics of sunflower as a rubber-producing plant[J]. Industrial Crops and Products，31（3）：481–491.

Perseguini J M K C，Romão L R C，Briñez B，et al.，2012. Genetic diversity of cultivated accessions and wild species of rubber tree using ESTSSR markers[J]. Pesquisa Agropecuária Brasileira，47（8）：1 087–1 094.

Ponciano G，McMahan C M，Xie W，et al.，2012. Transcriptome and gene expression analysis in cold-acclimated guayule（*Parthenium argentatum*）rubber-producing tissue[J]. Phytochemistry，79：57–66.

Post J，van Deenen N，Fricke J，et al.，2012. Laticifer-specific cis-prenyltransferase silencing affects the rubber，triterpene，and inulin content of *Taraxacum brevicorniculatum*[J]. Plant Physiology，158（3）：1 406–1 417.

Pramod S，Thomas V，Rao K S，2011. Structural and dimensional changes in the cambium of tapping panel dryness affected bark of *Hevea brasiliensis*[J]. Phyton-Annales Rei Botanicae，51（2）：231–244.

Régibeau P，Rockett K E，2016. Research assessment and recognized excellence：simple bibliometrics for more efficient academic research evaluations[J]. Economic Policy，31（88）：611–652.

Rivano F，Martinez M，Cevallos V，et al.，2010. Assessing resistance of rubber tree clones to *Microcyclus ulei* in large-scale clone trials in Ecuador: a less time-consuming field method[J]. European Journal of Plant Pathology, 126 (4): 541–552.

Rojruthai P, Sakdapipanich J T, Takahashi S, et al., 2010. In vitro synthesis of high molecular weight rubber by *Hevea* small rubber particles[J]. Journal of Bioscience and Bioengineering, 109 (2): 107–114.

Saha S K, Nair P K R, Nair V D, et al., 2010. Carbon storage in relation to soil size-fractions under tropical tree-based land-use systems[J]. Plant and Soil, 328 (1-2): 433–446.

Salvucci M E, Barta C, Byers J A, et al., 2010. Photosynthesis and assimilate partitioning between carbohydrates and isoprenoid products in vegetatively active and dormant guayule: physiological and environmental constraints on rubber accumulation in a semiarid shrub[J]. Physiologia Plantarum, 140 (4): 368–379.

Scott D J, da Costa B M T, Espy S C, et al., 2003. Activation and inhibition of rubber transferases by metal cofactors and pyrophosphate substrates[J]. Phytochemistry, 64 (1): 123–134.

Senevirathna A, Stirling C M, Rodrigo V H L, 2003. Growth, photosynthetic performance and shade adaptation of rubber (*Hevea brasiliensis*) grown in natural shade[J]. Tree Physiology, 23 (10): 705–712.

Shah A A, Hasan F, Shah Z, et al., 2012. Degradation of polyisoprene rubber by newly isolated bacillus sp. AF-666 from soil[J]. Applied Biochemistry and Microbiology, 48 (1): 37–42.

Singh A P, Wi S G, Chung G C, et al., 2003. The micromorphology and protein characterization of rubber particles in *Ficus carica*, *Ficus benghalensis* and *Hevea brasiliensis*[J]. Journal of Experimental Botany, 54 (384): 985–992.

Sookmark U, Pujade-Renaud V, Chrestin H, et al., 2002. Characterization of polypeptides accumulated in the latex cytosol of rubber trees affected by the tapping panel dryness syndrome[J]. Plant and Cell Physiology, 43 (11): 1 323–1 333.

Suzuki N, Uefuji H, Nishikawa T, et al., 2012. Construction and analysis of EST libraries of the trans-polyisoprene producing plant, *Eucommia ulmoides* Oliver[J]. Planta, 236 (5): 1 405–1 417.

Takaya A, Zhang Y W, Asawatreratanakul K, et al., 2003. Cloning, expression and characterization of a functional cDNA clone encoding geranylgeranyl diphosphate synthase of *Hevea brasiliensis*[J]. Biochimica Biophysica Acta, 1625 (2): 214–220.

Tang C, Huang D, Yang J, et al., 2010. The sucrose transporter *HbSUT3* plays an active role in sucrose loading to laticifer and rubber productivity in exploited trees of *Hevea brasiliensis* (para rubber tree) [J]. Plant, Cell and Environment, 33 (10): 1 708–1 720.

Tian W M, Shi M J, Yu F Y, et al., 2003. Localized effects of mechanical wounding an exogenous jasmonic acid on the induction of secondary laticifer differentiation in relation to the distribution of jasmonic acid in *Hevea brasiliensis*[J]. Acta Botanica Sinica, 45 (11): 1 366–1 372.

Tian W W, Huang W F, Zhao Y, 2010. Cloning and characterization of *HbJAZ1* from the laticifer cells in rubber tree (*Hevea brasiliensis* Muell. Arg.) [J]. Trees, 24 (4): 771–779.

Tungngoen K, Viboonjun U, Kongsawadworakul P, et al., 2011. Hormonal treatment of the bark of rubber trees (*Hevea brasiliensis*) increases latex yield through latex dilution in relation with the differential expression of two aquaporin genes[J]. Journal of Plant Physiology, 168 (3): 253–262.

Tomazic-Jezic V J, Woolhiser M R, Beezhold D H, 2002. ELISA inhibition assay for the quantitation of antigenic protein in natural rubber latex[J]. Journal of Immunoassay and Immunochemistry, 23 (3): 261–278.

van Deenen N, Bachmann A L, Schmidt T, et al., 2012. Molecular cloning of mevalonate pathway genes from *Taraxacum brevicorniculatum* and functional characterisation of the key enzyme 3-hydroxy-3-methylglutaryl-coenzyme a reductase[J]. Molecular Biology Reports, 39 (4): 4 337–4 349.

Watcharakul S, Umsakul K, Hodgson B, et al., 2012. Biodegradation of a blended starch/natural rubber foam biopolymer and rubber gloves by *Streptomyces coelicolor* CH13[J]. Electronic Journal of Biotechnology, 15 (1): 1–8.

Wititsuwannakul D, Chareonthiphakorn N, Pace M, et al., 2002. Polyphenol oxidases from latex of *Hevea brasiliensis*: purification and characterization[J]. Phytochemistry, 61 (2): 115–121.

Yang Z P, Li H L, Guo D, et al., 2012. Molecular characterization of a novel 14-3-3 protein gene (*Hb14-3-3c*) from *Hevea brasiliensis*[J]. Molecular Biology Reports, 39 (4): 4 491–4 497.

Zhang H, Zhang G L, 2003. Microbial biomass carbon and total organic carbon of soils as affected by rubber cultivation[J]. Pedosphere, 13 (4): 353–357.

Zhu J H, Chen X, Chang W J, et al., 2010. Molecular characterization of *HbCDPK1*, an ethephon-induced calcium-dependent protein kinase gene of *Hevea brasiliensis*[J]. Bioscience, Biotechnology, and Biochemistry, 74 (11): 2 183–2 188.

Zhang M, Zou X, Schaefer D A, 2010. Alteration of soil labile organic carbon by invasive earthworms（*Pontoscolex corethrurus*）in tropical rubber plantations[J]. European Journal of Soil Biology，46（2）：74–79.

Zhang Q，Zhu J，Ni Y，et al.，2012. Expression profiling of *HbWRKY1*，an ethephon-induced *WRKY* gene in latex from *Hevea brasiliensis* in responding to wounding and drought[J]. Trees，26（2）：587–595.

6 国内天然橡胶研究热点分析

6.1 引言

我国的天然橡胶种植主要集中在云南、海南和广东3个省份，云南与海南是我国橡胶主产区。在我国天然橡胶产业发展历程中，主要有4方面的成就，一是打破禁区，实现北纬18°～24°大面积成胶；二是形成600多万人口赖以生存的新型社区；三是打破当地刀耕火种的不良习惯；四是保障中国经济发展及政治调控对橡胶资源的需要。科学技术创新是推动产业发展的动力，我国一些重要科技成果加快了天然橡胶产业的发展，达到国际领先和国际先进水平。例如，建立了橡胶树籽苗嫁接及试管芽条繁殖技术体系，完成促进剂清洁生产工艺、橡胶助剂废水资源化成果，研发环保型脱蛋白恒门尼黏度天然橡胶制备技术等（国家天然橡胶产业技术体系，2015）。因此，在前期解析国际天然橡胶研究热点的基础上，对国内天然橡胶研究领域的科研成果进行全面分析，有利于发现国内天然橡胶领域的重要科研成果。

科学知识图谱是在科学计量与文献计量的理论、方法支撑下，通过对科技文本的可视化的技术来展示研究主题知识分布、结构和关联的新兴研究方向。随着数据科学和计算机科学的快速发展，科学知识图谱的方法也得到了快速发展和传播（李杰，2018；赵海莉等，2020；武丽丽等，2019）。本章以CNKI的期刊论文为数据源，利用VOSviewer技术，绘制国内天然橡胶研究领域的科学知识图谱。通过分析数据库中文献的关键词共现和高频关键词，以可视化的方式展现国内天然橡胶研究领域主题的整体分布和时序演变，明确该领域的研究热点及演化趋势，以期为我国天然橡胶的后续研究提供参考。

6.2 数据来源与研究方法

6.2.1 数据来源

本章基于CNKI数据库中的期刊论文，期刊类别包括SCI来源期刊、EI来源期刊、核心期刊、CSSCI和CSCD。时间跨度为2005—2019年，以SU=（“橡胶树”+“巴西橡胶”+“天然橡胶”+“天然胶乳”+“橡胶胶乳”+“橡胶草”+“产胶植物”+“银胶菊”+“银色橡胶菊”+“杜仲胶”+“橡胶林”+“橡胶人工林”+“橡胶种植园”+“橡胶园”+“植胶区”）+[“橡胶”*（“排胶”+“产胶”+“割胶”）]作为主题词进行相关论文检索。获取的论文数量为9 235篇（检索日期2020-03-26）。

6.2.2 研究方法

采用VOSviewer进行数据的可视化分析，VOSviewer（Visualization of Similarities Viewer，简称VOSviewer）是由Nees Jan van Eck和LudoWaltman联合开发的文献计量和可视化分析工具。目前该软件具备了几乎所有常见的文献计量分析功能，如文献耦合、共被引、合作以及共词分析等。主要特点在于以网络聚类图、密度图等来作为可视化结果，并可以有效地进行主题挖掘、文献共被引聚类等分析，已经广泛应用于多个学科领域的科学计量分析中（李杰，2018）。本章主要运用VOSviewer构建不同时期（2005—2009年，2010—2014年和2015—2019年）关键词的聚类视图和密度视图，同时统计各聚类关键词和高频关键词，结合分析2005—2019年国内天然橡胶领域的研究热点。

6.3 2005—2009年天然橡胶研究热点分析

科学论文中的关键词是作者根据论文内容精选的论文主题标引，是对所发表论文内容的高度凝练。因此，在科学论文的主题挖掘和热点分析中，关键词常被选择为分析对象。构建2005—2009年天然橡胶领域关键词共词聚类网络（图6-1a）。共词网络包含284个关键词，每个节点代表一个关键词，节点和标签的大小反映了关键词共现频次的高低；节点之间的连线表示共现关系，连线粗细代表共现关系的强弱。关键词由于共现强度的不同会形成集聚现象，并形成特定的主题聚类。2005—2009年天然橡胶研究的5个子领域被聚集在一起，并通过不同颜色和建议标签进行区分。在密度视图中（图6-1b），系统提供3种颜色分布，红色区域关键词出现的频次较黄色区域的高，绿色区域关键词出现的频次较黄色区域的低，可以直观地看到密度视图中重要的区域；关键词空间的距离分布表示他们之间的相似度。可以看出天然橡胶、天然胶乳、碳纳米管、改性、性能、纳米复合材料、橡胶树、橡胶园、生长等关键词在密度视图中呈现红色和黄色，这些词出现的频次较高。本章结合在VOSviewer中检测到聚类中引用频次较高的关键词来解析天然橡胶领域不同阶段的研究热点（表6-1）。

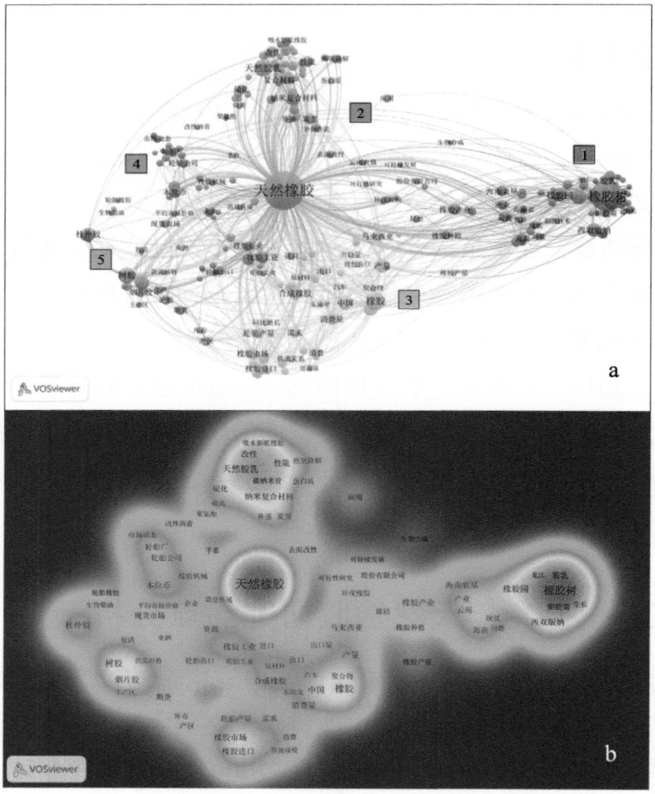

图6-1　2004—2009年国内天然橡胶研究关键词聚类视图（a）和密度视图（b）

Fig. 6-1　2004–2009 keyword cluster and density view in Chinese natural rubber research.（a）Cluster view；（b）Density view

表6–1 国内天然橡胶聚类高频关键词（2004—2009年）

Tab. 6–1 High frequency keywords during 2004–2009 in the cluster of Chinese natural rubber research

聚类编号 Cluster ID	高频关键词 High frequency keywords
#1	橡胶树；园艺作物；橡胶园；橡胶产业；云南；西双版纳；农场；胶乳；干胶含量；植胶区；海南；海南农垦；播种面积；橡胶种植；种植面积；生物合成；生物量；生长量；生物多样性；土壤养分；胶园土壤；橡胶林；茉莉酸；乙烯；有机质；养分
#2	天然橡胶；天然胶乳；力学性能；纳米复合材料；炭黑；补强剂；改性；复合材料；烯烃；性能；白炭黑；物理机械性能；丁苯橡胶；顺丁橡胶；硫化特性；共混；接枝；有机蒙脱土；环氧化天然橡胶；促进剂；胶料；马来酸酐；硫化；动态力学性能
#3	橡胶；弹性体；中国；消费量；产量；合成橡胶；橡胶进口；橡胶市场；轮胎产量；橡胶价格；马来西亚；同期；消费；合成材料；泰国；生产关系；合成弹性体；需求；供需；供求；聚合物；出口；市场价格；印度尼西亚
#4	本位币；美元；橡胶工业；子午线轮胎；橡胶行业；轮胎厂；轮胎公司；企业；企业管理；固特异；米其林；轮胎行业；轮胎出口；橡胶企业；市场动态；橡胶公司；轮胎企业；财政管理；普利司通公司；北美洲；平均市场价格；美国；行业动态
#5	标准胶；烟片胶；杜仲胶；产区；现货市场；沪胶；古塔波胶；资源；期货；库存；期货市场；金融市场；杜仲；牛市；亚洲；均价；供需形势

聚类#1讨论中国植胶区橡胶种植面积、胶乳产量、橡胶生物合成及橡胶人工林生态系统土壤养分、生物量与生物多样性等（红色）。关键词有中国、云南、海南、资源、种植面积、橡胶树、胶乳、乙烯、产量、生物量、茉莉酸、生物合成、橡胶人工林、养分、有机质、生物多样性、热带雨林季节等。例如，邹智等介绍橡胶树橡胶生物合成的分子过程、参与合成的主要酶类和辅助因子，以及割胶、茉莉酸和乙烯等调控橡胶合成。贾开心等测定西双版纳3个海拔梯度上橡胶林地上及各器官的生物量；庞家平等对橡胶—大叶千斤拔复合生态系统和橡胶纯林进行连续3年的土壤养分对比观测（邹智等，2009；贾开心等，2006；庞家平等，2009）。

聚类#2关注在天然橡胶复合材料改性与性能研究，重点关注炭黑、白炭黑、马来酸酐、碳纳米管等改性天然橡胶或天然橡胶/丁苯橡胶或顺丁橡胶或环氧化天然橡胶等及其复合材料的力学、物理机械性能和硫化特性研究（绿色）。例如，隋刚等对碳纳米管/天然橡胶复合材料的制备工艺和材料性能进行了研究；吕明哲等介绍了利用动态机械热分析仪（DMA）研究天然橡胶及其改性物的低温动态力学性能；刘吉文等实现了环氧天然橡胶（ENR）对白炭黑的固态原位接枝，探讨了白炭黑和ENR的反应配比对增强性能的影响（隋刚等，2005；吕明哲等，2007；刘吉文等，2008）。

聚类#3集中在合成橡胶价格与供求关系方面，由于我国在2004年后取消了天然橡胶配额管理，采取天然橡胶进口等级制度，天然橡胶进口从此进入了更加市场化的运作模式。关键词如合成橡胶、橡胶进口、橡胶价格、橡胶市场、消费、供需等（黄色）。例如，梁诚介绍了热塑性弹性体特性及应用、生产现状、市场消费和发展趋势；高云芝等介绍国内对溴化丁基橡胶的应用、性能的研究及国内外卤化丁基橡胶的生产消费情况等；刘川介绍了世界异戊橡胶生产及消费现状（梁诚，2005；高云芝等，2007；刘川，2009）。

聚类#4集中在橡胶轮胎行业动态，聚氨酯橡胶复合材料制备及其硫化特性方面，关键词包括子午线轮胎、轮胎公司/企业、固特异、米其林、轮胎出口、市场动态等（蓝色）。马良清等概述汽车轮胎行业特点及影响轮胎工业发展的有利因素和制约轮胎工业发展的不利因素，进一步对新时期轮胎市场出现的新问题和国内外轮胎产品标准的差异等进行阐述；胡工分析轮胎行业的生产、销售及出口情况，部分轮胎企业的投资情况和产量规模，分析了与之相关企业的现状；蔡为民根据我国轮胎产品出口受到国际需求疲软的影响提出应对措施和建议（马良清等，2009；胡工，2007；蔡为民，2009）。

聚类#5是较小的聚类，关注天然橡胶现货、期货价格，橡胶出口与供求关系和杜仲胶的开发（紫色）。高勇等运用协整序列分解模型和数据选取方法，发现不同的天然橡胶期、现货价格序列均存在显著的协整关系；曹旭平等对1998—2007年中国天然橡胶的产地分布、供需矛盾、进口市场及进口品种结构进行了实证分析；尤飞通过分析发现，全球天然橡胶消费重心呈现梯度转移格局，全球天然橡胶产业有阶段性空间转移规律，其消费与化工行业的发展有着密切的联系（高勇，2008；曹旭平和沈杰，2009；尤飞，2005）。

6.4　2010—2014年天然橡胶研究热点分析

2010—2014年共词网络包括247个关键词（图6-2a），形成5个聚类，即聚类#1天然橡胶复合材料力学、物理、机械性能增强及其结构特征（红色）；聚类#2橡胶树胶乳基因表达分析（绿色）；聚类#3橡胶林生态系统土壤养分和生物多样性（蓝色）；聚类#4杜仲胶的应用（黄色）；聚类#5轮胎用橡胶的混炼工艺和性能（紫色）。表6-2列出5个聚类中的主要关键词。

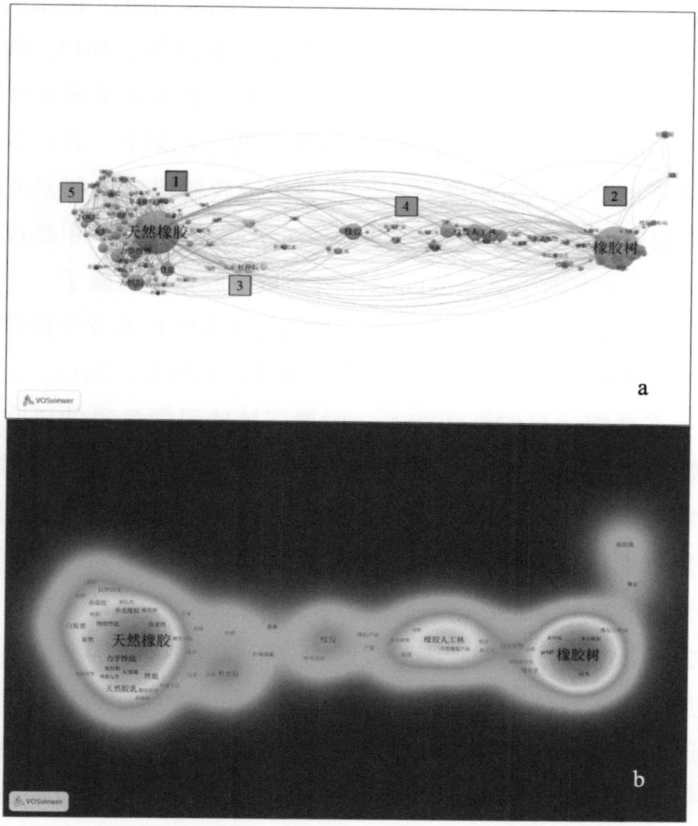

图6-2　2010—2014年国内天然橡胶研究关键词聚类视图（a）和密度视图（b）

Fig. 6-2　2010-2014 keyword cluster and density view in Chinese natural rubber research.（a）Cluster view；（b）Density view

天然橡胶领域关键词聚类分布紧凑，但分布不均衡，其中聚类#1、#2和#3所占比例较大，反映了2010—2014年天然橡胶研究的重点。值得注意的是，在#1中出现频率较高的词有“天然橡胶”“力学性能”“复合材料”等，并且围绕它们形成了一个较大的网络。实际上，在天然橡胶复合材料研究中，天然橡胶的化学改性、共混改性、纳米改性、结构—性能机理等研究，通常对评估复合材料的力学或物理性能至关重要。因此，天然胶乳填料改性（纳米粒子和非纳米粒子改性）及化学改性（氯化、环化、环氧化、接枝聚合）是该领域研究的重点；天然橡胶的环氧化改性，环氧化天然橡胶（ENR）制备和表征方法、ENR与白炭黑之间的相互作用、ENR的硫化及其在轮胎胶料中的应用等也是该领域研究的热点（彭政等，2014；何灿忠等，2012；刘东辉等，2012）。#2也是较大的聚类，主要关键词有橡胶树、胶乳、橡胶草、基因、克隆、表达分析、序列分析、乙烯利、死皮等。例如，庄海燕等从不同学科不同角度对乙烯利刺激橡胶树增产机制研究的结果进行分析；王启超等通过比较甲羟戊酸途径关键酶3-羟基-3-甲基戊二酸单酰辅酶A还原酶（HMGR）氨基酸同源区域，从橡胶草中克隆了一个HMGR基因；邹智等讨论了橡胶树不同“死皮”类型的主要特征和发生规律，并提出防控策略（庄海燕等，2010；王启超等，2012；邹智等，2012）。#3涉及橡胶人工林生态系统，与#2密切相关，同属于植物科学学科的研究重点。例如，刘晓娜等基于Landsat数据和MODIS-NDVI数据，采用决策树分类的方法提取中老缅交界地区的橡胶林地；方丽娜等通过去除凋落物和切根控制试验，研究热带森林不同土地利用方式（次生林/橡胶林）对西双版纳热带森林土壤微生物生物量碳的影响（刘晓娜等，2013；方丽娜等，2011）。

图6-2b显示了基于密度视图的关键词分析，其中颜色从红色到绿色的演变代表了热门主题随时间的变化。可以看出，在聚类#1中可以找到近期的研究主题，这表明目前天然橡胶研究的主要关注点。特别是聚类#1通过对天然橡胶进行化学、共混、纳米改性等制备复合材料是研究的热点，越来越受到人们的关注。聚类#2和聚类#3是橡胶树胶乳相关基因表达分析、橡胶人工林生态系统等方面的研究课题，也在天然橡胶研究中占有较大的比重，是目前正在进行的研究，在今后的工作中会继续受到关注。

表6-2 国内天然橡胶聚类高频关键词（2010—2014年）

Tab. 6-2 High frequency keywords during 2010-2014 in the cluster of Chinese natural rubber research

聚类编号 Cluster ID	高频关键词 High frequency keywords
#1	天然橡胶；力学性能；复合材料；白炭黑；性能；天然胶乳；炭黑；硫化特性；环氧化天然橡胶；物理性能；胎面胶；交联密度；改性；异戊橡胶；动态力学性能；物理机械性能；纳米复合材料；丁苯橡胶；补强；硫化胶；耐磨性能；顺丁橡胶；动态性能；石墨烯；并用胶；热导率；淀粉；有机蒙脱土；蒙脱土；加工性能；老化
#2	橡胶树；基因克隆；表达分析；胶乳；序列分析；死皮；基因表达；克隆；橡胶草；乙烯利；寒害；气刺微割；砧木；银胶菊；遗传转化；炭疽病；鉴定；原核表达；杀菌剂；质谱；生物信息学；白粉病；接穗；棒孢霉落叶病；茉莉酸；表达；RT-PCR；品种
#3	橡胶人工林；橡胶园；西双版纳；园艺作物；云南；干胶含量；产量；现状；间作；割胶技术；橡胶种植；监测；遥感；相关性；空间变异；土壤有机碳；聚异戊二烯；表征；GIS；土壤；胶农；土壤肥力

6.5 2015—2019年天然橡胶研究热点分析

2015—2019年共词网络包括272个关键词（图6-3a），聚类#1关注改性天然橡胶制备纳米复合材料及其性能表征（红色），包含115个关键词。聚类#2标记为橡胶树胶乳相关基因的克隆和表达（绿色），包括51个关键词。聚类#3涉及土地利用变化下橡胶林土壤养分变化（蓝色），包括42个关键词。聚类#4至聚类#6是较小的聚类（黄色、紫色和天蓝），包括64个关键词，内容涉及填充橡胶的Payne效应研究、异戊二烯共聚橡胶的结构与组成表征、“一带一路”背景下天然橡胶产业发展形势等（陈明文，2016；王浩等，2015；刘涛等，2015）。聚类#2和聚类#3在图中位置接近，反映出它们的工作比聚类#1中的工作更紧密地联系在一起。以下主要对前3个聚类进行简要解释（图6-3，表6-3）。

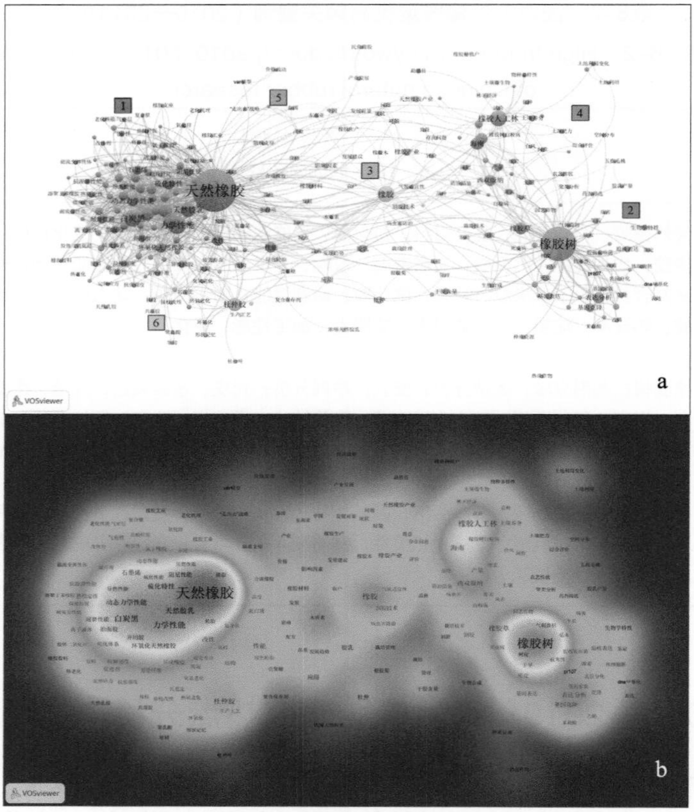

图6-3　2015—2019年国内天然橡胶研究关键词聚类视图（a）和密度视图（b）

Fig. 6-3　2015-2019 keyword cluster and density view in Chinese natural rubber research.（a）Cluster view;（b）Density view

图6-3a红色部分所示，该聚类关注的是天然橡胶纳米复合材料制备及其性能表征。侧重于改性炭黑或白炭黑或石墨烯或氧化石墨烯等补强天然橡胶制备纳米复合材料研究，具体方法包括离子液体改性、硅烷偶联剂改性、改性剂改性、机械共混、乳液共混、原位还原、湿法混炼等（刘尧华等，2016；穆晓东等，2017；王经逸等，2016；汤银银等，2015）。对复合材料的结构（分散性、相容性、交联密度、微观形貌等）、性能（物理、机械、力学、导热、耐磨性等）和加工特性等的影响研究也是重点，如改性白炭黑在天然橡胶基体中的分散性明显提高；改性纳米SiO_2与复合材料的相容性明显提高，大大缩短胶料的正硫化时间，增大总交联密度；同时提高

SiO_2-HBP-L/SBR纳米复合材料的力学和耐磨性能；通过胶乳—双辊连用法制备炭黑/天然橡胶复合材料（RCB）、碳纳米管/天然橡胶复合材料（RCNT）和石墨烯/天然橡胶复合材料（RGE）。该方法可以将填料均匀分散在橡胶基体中，Payne效应增强（崔凌峰等，2017；战艳虎等，2017；张颖等，2016）。该聚类具有很强的跨学科性质，包括高分子科学、材料科学、工程化学和物理化学等。图6-3a绿色部分所示，聚类#2关注橡胶树相关基因克隆与表达分析，与聚类#1相比，它也注重学科，具有更多的学科重点，如植物科学、基因工程、生物化学与分子生物学等。过程参数设置和优化的试验分析是本聚类的研究重点，旨在建立和讨论与非生物胁迫或外界刺激下橡胶树或胶乳基因克隆与表达相关的知识。聚类包括的分子生物学术语，如克隆、转录因子、延伸因子、基因、表达分析，以及与试验过程相关的术语，如实时荧光定量PCR、RT-PCR、RACE技术、同源克隆、Blast分析等。聚类中出现了各种外界环境因素，如干旱、低温、伤害、割胶、非生物胁迫、乙烯利、茉莉酸等（张凤良等，2016；位明明等，2016；刘辉等，2015；赵武帅等，2015；赵丽娟等，2015）。该聚类与聚类#3土地利用变化下橡胶林土壤养分变化在学科上有联系。

图6-3a蓝色部分所示，这一聚类重点关注与橡胶林生态环境相关的研究。聚类#3关注生态学、土壤学、园艺学等学科。从土地利用空间分布特征、时空变化规律、土地利用程度对橡胶园、茶园、林地、灌木林、草地等土地利用变化特征进行分析是这一聚类的重要论述，以建立和讨论关于土地利用变化背景下橡胶林土壤养分变化的知识。这些术语集中于所研究的指标类型，例如有机质、全氮、速效磷、速效钾、交换性酸、交换性盐基离子、pH值、碱解氮、铵态氮、硝态氮、微生物生物量碳等。不同的技术方法开展野外或田间试验是研究的重点，包括土壤理化性质测定、凯氏定氮法、氯仿熏蒸浸提法等；同时也包括建模工作和数值分析，如GIS技术、数字土壤制图、信息挖掘等（陈玉芹等，2019；陈永川等，2019；孙海东等，2016；廖谌婳等，2015；郭澎涛等，2015）。热力视图同样反映了研究热点的变化（图6-3b），即改性天然橡胶纳米复合材料制备及其性能表征，橡胶树相关基因克隆与表达分析及橡胶林生态环境研究。

表6-3　国内天然橡胶聚类高频关键词（2015—2019年）

Tab. 6-3　High frequency keywords during 2015-2019 in the cluster of Chinese natural rubber research

聚类编号 Cluster ID	高频关键词 High frequency keywords
#1	天然橡胶；白炭黑；力学性能；复合材料；炭黑；物理性能；动态力学性能；硫化特性；天然胶乳；改性；顺丁橡胶；物理机械性能；石墨烯；加工性能；耐磨性能；阻尼性能；丁苯橡胶；并用胶；硅烷偶联剂；环氧化天然橡胶；丁腈橡胶；湿法混炼；碳纳米管；导热性能；分散性；氧化石墨烯；滚动阻力；门尼黏度；胎面胶；交联密度；促进剂；偶联剂；补强剂；抗湿滑性能；离子液体；表面改性；硫化体系；异戊橡胶；拉伸强度；三元乙丙橡胶；混炼工艺；热稳定性；Payne效应；溴化丁基橡胶；气密性；氯丁橡胶；微观形貌；碳纤维；共混
#2	橡胶树；表达分析；基因克隆；西双版纳；基因表达；胶乳；死皮；割胶；克隆；生物信息学；乙烯利；基因家族；干旱；干胶含量；生物合成；原核表达；生物学特性；生长；非生物胁迫；乙烯；茉莉酸；生理参数；低温；割龄；园艺作物；气候适宜性；激素；乳管细胞
#3	橡胶人工林；海南；橡胶园；橡胶草；产量；云南；寒害；土壤；施肥；土壤养分；农艺性状；间作；风险区划；土壤肥力；综合评价；聚类分析；土地利用；物种多样性；生物量；空间分布；台风；品种；土地利用变化；林龄；涡度相关；相关性；套种；遥感；风害；VAR模型

6.6　2005—2019年天然橡胶研究热点演化分析

将2005—2019年天然橡胶研究论文分为2005—2009年、2010—2014年和2015—2019年3个阶段。依次对3个阶段的文献进行关键词聚类分析，得出国内的天然橡胶热点演变情况。总体来说，在2005—2009年阶段研究热点较分散，涉及橡胶种植面积、胶乳产量、橡胶生物合成，橡胶人工林生态系统土壤养分、生物量与生物多样性，天然橡胶复合材料改性与性能以及天然橡胶现货和期货价格，合成橡胶价格、橡胶出口与供求关系，橡胶轮胎行业动态等方面。2010—2014年在橡胶树胶乳产量方面注重乙烯利刺激增产机制研究；橡胶人工林生态系统研究侧重于橡胶林遥感识别和影像；在天然橡胶复合材料研究基础上扩展了天然橡胶的化学改性、共混改性、纳米改性、结构—性能机理等；新增研究热点是天然橡胶的环氧化改性，环氧化天然橡胶（ENR）制备和表征方法、ENR的硫化及其在轮胎胶料中的应用。2015—2019年研究热点的突出变化是在材料科学领域，通过离子液体改性、硅烷偶联剂改性、改性剂改性等补强天然橡胶制备纳米复合材料及其对复合材料的结构（分散性、相容性、交联密度、微观形貌等）、性能（物理、机械、力学、导热、耐磨性等）和加工特性等的影响；在植物科学领域，关注非生物胁迫或外界刺激下橡胶树或胶乳基因克隆与表达分析；在生态学学科领域，集中在土地利用变化背景下橡胶林土壤养分变化分析。整体上看国内研究热点演变较快，趋于集中在材料科学、植物科学和生态学三大学科领域。

6.7 本章小结

天然橡胶领域交叉学科研究明显，仍处于较快发展阶段。研究热点的整体分布表明，国内天然橡胶研究的三大核心热点如下：

（1）橡胶轮胎和非轮胎橡胶制品及产业链上下游，以及橡胶循环利用，具体表现在填料改性和化学改性补强天然橡胶制备纳米复合材料及其对复合材料的结构、性能和加工特性等的影响。

（2）非生物胁迫或外界刺激下橡胶树或胶乳基因克隆与表达分析，这为橡胶树的高产、抗病、抗旱、抗寒等重要经济性状的基因组选择育种与优异基因资源的发掘利用提供了高质量橡胶树参考基因。

（3）土地利用变化背景下橡胶人工林土壤养分变化及生物多样性，这是在全球经济下行、中美贸易摩擦加剧、环保治理形势更加严峻的大背景下形成的研究热点。国内研究热点演变较快，研究内容集中在三大学科领域，即化学与材料科学、生物科学及生态与环境科学。

参考文献

蔡为民，2009. 应对我国轮胎出口贸易摩擦的措施和建议[J]. 轮胎工业，29（12）：716-718.

曹旭平，沈杰，2009. 中国天然橡胶供需矛盾及进口结构演变分析[J]. 林业经济问题，29（2）：153-157.

崔凌峰，熊玉竹，戴骏，等，2017. 改性白炭黑/天然橡胶复合材料的制备及性能[J]. 高分子材料科学与工程，33（5）：158-163.

陈明文，2016. 我国天然橡胶产业发展形势与因应策略[J]. 农业经济问题，37（10）：91-94，112.

陈永川，刘忠妹，许木果，等，2019. 西双版纳橡胶林土壤氮的分布特征及与橡胶树生长的关系[J]. 西南农业学报，32（3）：584-589.

陈玉芹，胡永亮，张丽萍，等，2019. 基于主成分和聚类分析的德宏橡胶林土壤肥力评价[J].热带作物学报，40（8）：1 461-1 467.

方丽娜，杨效东，杜杰，2011. 土地利用方式对西双版纳热带森林土壤微生物生物量碳的影响[J]. 应用生态学报，22（4）：837-844.

高勇，郭彦峰，2008. 中国天然橡胶期货的套期保值比率与绩效研究[J]. 工业技术经济（7）：158-161.

高云芝，田恒水，张新军，等，2007. 溴化丁基橡胶的应用研究及市场分析[J]. 橡胶科技市场（2）：4-7.

郭澎涛，李茂芬，罗微，等，2015. 基于多源环境变量和随机森林的橡胶园土壤全氮含量预测[J]. 农业工程学报，31（5）：194-200，201，202.

国家天然橡胶产业技术体系，2016. 中国现代农业产业可持续发展战略研究天然橡胶分册[M]. 北京：中国农业出版社. 11-35.

何灿忠，彭政，钟杰平，等，2012. 环氧化天然橡胶的研究进展[J]. 高分子通报（2）：84-93.

贾开心，郑征，张一平，2006. 西双版纳橡胶林生物量随海拔梯度的变化[J]. 生态学杂志，25（9）：1 028-1 032.

胡工，2007. 我国轮胎行业发展现状及趋势[J]. 中国石油和化工经济分析（7）：57-62.

李杰，2018. 科学计量与知识网络分析：方法与实践[M]. 第二版. 北京：首都经济贸易大学出版社. 332-337.

梁诚，2005. 热塑性弹性体生产现状与发展趋势[J]. 石油化工技术经济（1）：35-40.

廖谌婳，封志明，李鹏，等，2015. 中老缅泰交界地区土地利用变化信息挖掘与国别对比[J].自然资源学报，30（11）：1 785-1 797.

刘川，2009. 世界异戊橡胶生产现状及我国发展前景[J]. 橡胶科技市场，7（12）：1-5.

刘东辉，徐亚丽，孟丽丰，2012. 天然胶乳改性研究进展[J]. 应用化工，41（1）：158-163.

刘辉，邓治，陈江淑，等，2015. 巴西橡胶树类钙调素蛋白基因*HbCML27*克隆与表达分析[J].分子植物育种，13（12）：2 721-2 727.

刘吉文，许海燕，吴驰飞，2008. 环氧天然橡胶接枝高分散白炭黑增强天然橡胶复合材料的制备及表征[J]. 高分子学报（2）：123-128.

刘尧华，林宇，张栋葛，等，2016. 天然橡胶/石墨烯纳米复合材料的制备及耐核辐射性能[J].高等学校化学学报，37（7）：1 402-1 407.

刘涛，陈亚薇，刘东，等，2015. 填充橡胶的Payne效应[J]. 特种橡胶制品，36（6）：76-81.

刘晓娜，封志明，姜鲁光，2013. 基于决策树分类的橡胶林地遥感识别[J]. 农业工程学报，29（24）：163-172，365.

吕明哲，李普旺，黄茂芳，等，2007. 用动态热机械分析仪研究橡胶的低温动态力学性能[J].中国测试技术，33（3）：27-29.

马良清，王琰，孙宁，等，2009. 我国汽车轮胎行业现状及发展[J]. 轮胎工业，29（12）：707-715.

穆晓东，崔雨果，方庆红，等，2017. 白炭黑的功能化改性及其改性橡胶基复合材料的制备与表征[J]. 复合材料学报，34（1）：67-74.

庞家平，陈明勇，唐建维，等，2009. 橡胶—大叶千斤拔复合生态系统中的植物生长与土壤水分养分动态[J]. 山地学报，27（4）：433-441.

彭政，钟杰平，廖双泉，2014. 天然橡胶改性研究进展[J]. 高分子通报（5）：41-48.

隋刚，杨小平，梁吉，等，2005. 碳纳米管/天然橡胶复合材料的制备及性能[J]. 复合材料学报，22（5）：72-77.

孙海东，刘备，吴炳孙，等，2016. 橡胶树人工林地土壤酸度特征及酸化原因分析[J]. 西北林学院学报，31（2）：49-54.

汤银银，王娜，杨凤，等，2015. 机械共混法制备改性氧化石墨烯/天然橡胶复合材料及性能表征[J]. 高分子材料科学与工程，31（9）：167-172.

王浩，邹陈，贺爱华，2015. 高反式-1，4-丁二烯-异戊二烯共聚橡胶的结构表征及其在轿车轮胎子口护胶中的应用研究[J]. 高分子学报（12）：1 387-1 395.

王经逸，张旭敏，刘鹏章，等，2016. 离子液体改性氧化石墨烯对天然橡胶性能的影响Ⅰ. 物理机械性能和导热性能[J]. 合成橡胶工业，39（1）：10-14.

王启超，刘实忠，校现周，2012. 橡胶草HMGR基因的克隆及表达分析[J]. 植物研究，32（1）：61-68.

位明明，李维国，高新生，等，2016. 巴西橡胶树响应乙烯利刺激的生理及其分子调控机制研究进展[J]. 生物技术通报，32（3）：1-11.

武丽丽，孙爻，张礼生，等，2019. 基于CNKI文献计量的我国生物防治学科研究进展与发展态势分析[J]. 中国生物防治学报，35（6）：958-965.

尤飞，2005. 全球天然橡胶供需格局与产业形势分析[J]. 中国农垦（8）：16-18.

战艳虎，孟艳艳，夏和生，2017. 不同维数填料对橡胶Payne效应的影响[J]. 高分子材料科学与工程，33（1）：92-96.

张凤良，毛常丽，李小琴，等，2016. 橡胶树优树无性系对干旱胁迫的生理响应[J]. 西北林学院学报，31（4）：67-72.

张颖，彭健，林勇，等，2016. 长链超支化聚酯改性纳米SiO_2及其在丁苯橡胶中的应用[J]. 高分子学报（6）：706-714.

赵海莉，张婧，2020. 基于CiteSpace和VOSviewer的中国水旱灾害研究进展与热点分析[J].

生态学报，40（12）：1-10.

赵丽娟，袁彬青，王秀珍，等，2015. 橡胶延伸因子*REF*基因的克隆、转化及功能分析[J]. 西北农业学报，24（1）：144-150.

赵武帅，翟琪麟，安泽伟，等，2015. 橡胶树转录因子HbWRKY9的克隆与特性分析[J]. 基因组学与应用生物学，34（3）：599-606.

庄海燕，安锋，张硕新，等，2010. 乙烯利刺激橡胶树增产机制研究进展[J]. 林业科学，46（4）：120-125.

邹智，杨礼富，王真辉，等，2009. 橡胶树中橡胶的生物合成与调控[J]. 植物生理学通讯，45（12）：1 231-1 238.

邹智，杨礼富，王真辉，等，2012. 橡胶树“死皮”及其防控策略探讨[J]. 生物技术通报（9）：8-15.

7 国际天然橡胶研究前沿及其演进历程分析

7.1 引言

橡胶是热带地区典型的经济林作物，是重要的人工林生态系统，在热带农林业中具有特殊和重要地位。与此同时，在军工、高端和特殊用途橡胶领域中，天然橡胶仍具有不可替代性。当前，天然橡胶价格持续低迷，天然橡胶制品长期处于供大于求的状态，但国产胶自给率仍不到20%。而长期以来，高端和特殊用途的高性能胶却几乎完全依赖进口。我国天然橡胶产业发展中重大问题已经从早期的追求高产转变为高产与优质并重。在我国天然橡胶产业的发展历程中，科学技术对于促进产业发展和产业升级具有重要意义。一些重要技术引进、革新和应用极大地促进了产业的发展，如一系列高产抗性品种的引种、选育和推广，乙烯利刺激的低频割胶技术等。在天然橡胶产业发展的新形势下，追踪和把握国际天然橡胶科技前沿，对于解决我国天然橡胶产业中重大科学和需求问题，加快实现我国天然橡胶产业升级和可持续发展具有重要的参考价值。

本章是在第5章国际天然橡胶研究热点分析的基础上，进一步分析国际天然橡胶领域的研究前沿及其演进历程。主要利用科学计量方法结合CiteSpace技术，对1988—2017年Web of Science收录的天然橡胶文献进行共被引和聚类分析，以可视化图谱方式呈现国际天然橡胶领域的研究前沿、演进历程和演进趋势，并识别重要里程碑文献。

7.2 数据来源与研究方法

7.2.1 数据来源

数据来源于Web of Science核心合集的Science Citation Index Expanded（SCI-E）和Social Sciences Citation Index（SSCI）数据库。本章定义的天然橡胶是指从橡胶树（*Hevea brasiliensis*）、银胶菊（*Parthenium argentatum*）、蒲公英（*Taraxacum brevicorniculatum*）等提取的天然橡胶；本章的天然橡胶文献包括天然橡胶在植物科学和加工领域的相关研究。采用表7-1制定的检索策略进行主题检索，时间跨度为1988—2017年，文献类型为“Article”或“Review”，语种选择“English”，获得12 907篇文献，包含234 221篇引文（检索日期：2017-12-12，数据库更新日期：2017-12-11）。

表7-1　天然橡胶文献数据集检索策略

Tab. 7-1　Topic search queries used for data collection on natural rubber

检索式 Retrieval type	记录数 Records	检索策略 Topic search
#1	10 960	主题：（“rubber tree*” or “*Hevea*” or “natur* rubber” or “natur* latex” or “nr latex” or “rubber latex”）AND 文献类型：（Article OR Review）AND 语种：（English） 索引=SCI-EXPANDED，SSCI 时间跨度=1988–2017
#2	4 361	主题：（“rubber”）AND主题：（“tapping” or “plantation*” or “yard” or “garden” or “forest” or “NR” or “planting area*” or “planting” or “growing area*” or “growing state”）AND 文献类型：（Article OR Review）AND 语种：（English） 索引=SCI-EXPANDED，SSCI 时间跨度=1988–2017
#3	12 907	#1 OR #2 索引=SCI-EXPANDED，SSCI 时间跨度=1988–2017

7.2.2 研究方法

通过制定检索策略建立天然橡胶文献数据集，利用CiteSpace（5.0.R8.SE）进行数据集的可视化分析。CiteSpace是以一组文献记录作为输入，根据每年出版的文献形成一系列时间序列网络，并综合这些独立的网络，对基础领域的知识结构进行建模。CiteSpace支持多种类型的文献分析，包括合作网络分析、共词分析、作者共被引分析、文献共被引文分析等（Chen，2012a，2017；Chen et al.，2012b；Chen and Leydesdorff，2014）。

对1988—2017年天然橡胶文献进行共被引分析。首先，选取每一年中被引频次最高的前100篇引文，构建当年的共被引网络。然后，将各年的网络进行合成。合成的网络被划分为不同的聚类，与这些聚类相关联的施引文献代表了该领域的研究前沿，每个聚类中的共被引文献代表了知识基础。在生成共被引图谱之后，将引文发表年份作为x轴，聚类编号作为y轴，布局得到共被引网络的时间线图谱。时间线视图可以呈现各个聚类发展演变的时间跨度和研究进程。

7.3 天然橡胶文献共被引网络分析

图7-1表示1988—2017年每年引用次数排名前100的文献合成的网络图谱，合成网络中共有1 850篇引文，包含367个共被引网络聚类。3个最大的子网络有979篇引文，占整个网络的52%。网络具有非常高的模块度，为0.882 6，这表明天然橡胶各专业领域在共被引聚类中具有明确的界定。平均轮廓值为0.278 4，相对较低，是因为有大量的小聚类造成的。每个节点代表一篇在文献检索数据集中被引用过的参考文献，两节点之间的连线表示这两篇参考文献存在一次或多次共被引关系。共被引连线和聚类区块的颜色根据首次共被引的年份设定。蓝色连线或区块要比绿色连线或区块早，黄色要比绿色晚。每个聚类标签从引用该聚类的施引文献的标题、关键词和摘要进行提取和标记。

7.3.1 网络中高被引文献

被引次数高的文献通常被认为具有里程碑式的开创性贡献（表7-2）。聚类#0、#1和#3分别有3篇里程碑式的文献排在前10名，聚类#10有1篇文献。整个网络中被引次数最高的是Arroyo等（2003）的论文，引用191次，是关于有机改性蒙脱土替代炭黑制备天然橡胶纳米复合材料；第2是Lagier等（1992）的论文，引用115次，评估了手术室护士胶乳过敏的患病率；第3是Rattanasom等（2007）引用112次的文献，是关于二氧化硅和炭黑作为混合填料强化天然橡胶力学性能。排名第4和排名第10的文献来自聚类#0，排名第

5～7的文献均来自聚类#1，排名第8和排名第9的文献来自聚类#3。表明高被引文献关注胶乳过敏（#0、#1）和天然橡胶纳米复合材料（#3、#10）。

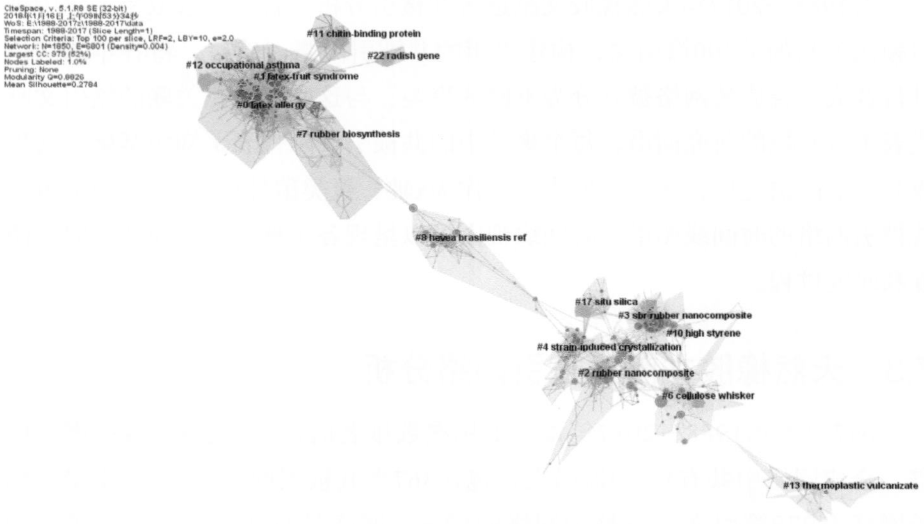

图7-1　1988—2017年天然橡胶文献共被引网络图谱

Fig. 7-1　Landscape view of the co-citation network on natural rubber literature during 1988-2017

表7-2　1988—2017年天然橡胶领域的高被引文献

Tab. 7-2　Most cited references of natural rubber domain during 1988-2017

排名 Rank	被引次数 Citation counts	被引文献（作者，年，期刊，卷，页） Cited references（Author，Year，Journal，Volume，Page）	聚类# Cluster #
1	191	Arroyo M，2003，Polymer，V44，P2447	3
2	115	Lagier F，1992，Journal of Allergy and Clinical Immunology，V90，P319	0
3	112	Rattanasom N，2007，Polymer Testing，V26，P369	10
4	102	Kelly K J，1993，Journal of Allergy and Clinical Immunology，V91，P1140	0
5	102	Blanco C，1994，Annals of Allergy，V73，P309	1
6	100	Czuppon A B，1993，Journal of Allergy and Clinical Immunology，V92，P690	1
7	97	Beezhold D H，1996，Clinical and Experimental Allergy，V26，P416	1
8	97	Varghese S，2003，Polymer，V44，P4921	3
9	95	Ray S S，2003，Progress in Polymer Science，V28，P1539	3
10	95	Turjanmaa K，1987，Contact Dermatitis，V17，P270	0

7.3.2　网络中高中介中心性文献

中介中心性测度的是节点在网络中位置的重要性。有两类节点可能具有较高的中介中心性：一是与其他节点高度相连的枢纽节点；二是位于不同聚类之间的节点。第二类节点更重要，它们比第一类节点更可能促使新兴趋势的出现（Chen，2012a，2017；Chen et al.，2012b；Chen and Leydesdorff，2014）。表7–3显示了网络中8篇结构上具有重要性的文献，这些文献不仅对它们如何连接网络中的单个节点，还对它们如何连接各节点的组合（如共被引聚类）来说是非常重要的。其中5个节点在聚类#8中，其他3个节点分别在聚类#7、聚类#1和聚类#2中。在本章定义的天然橡胶领域，这些文献可能对应的是变革性的科学发现。

表7–3　1988—2017年天然橡胶领域高中心度引文

Tab. 7–3　Cited citations with the highest betweenness centrality of natural rubber domain during 1988–2017

排名 Rank	中心度 Centrality	被引文献（作者，年，期刊，卷，页） Cited references（Author，Year，Journal，Volume，Page）	聚类# Cluster#
1	0.28	Ko J H，2003，Plant Molecular Biology，V53，P479	8
2	0.28	Tarachiwin L，2005，Biomacromolecules，V6，P1858	8
3	0.28	Oh S K，1999，Journal of Biological Chemistry，V274，P17132	7
4	0.26	van Beilen J，2007，Trends Biotechnol，V25，P522	8
5	0.24	Wagner B，1999，Journal of Allergy and Clinical Immunology，V104，P1084	1
6	0.22	Bokobza L，2007，Polymer，V48，P4907	2
7	0.15	Hillebrand A，2012，Plos One，V7，Pe41874	8
8	0.13	Wititsuwannakul R，2008，Phytochemistry，V69，P1111	8

7.3.3　网络中高突现性文献

引用突现性是指文献引用次数的激增，包括突发强度和突发状态持续时间两个属性（Chen，2012a，2017；Chen et al.，2012b；Chen and Leydesdorff，2014）。表7–4列出了1988—2017年天然橡胶文献数据集中具有高突现性的9篇引文。有3篇高突现性引文来自聚类#3，聚类#0和聚类#1各2篇，聚类#15和聚类#2各1篇。具有强烈突现性的引文也往往被视为产生了革命性的科学发现，如第一个革命性的科学发现文献是关于蒙脱土替代炭黑制备天然橡胶纳米复合材料（Arroyo et al，2003），下一个革命性的科学发现文献在胶乳过敏主题领域，关于医护人员发生胶乳过敏反应的概率（Turjanmaa，1987）。

表7–4　1988—2017年天然橡胶领域高突现性引文

Tab. 7–4　References with the strongest citation bursts of natural rubber domain during 1988–2017

排名 Rank	突现值 Citation bursts	被引文献（作者，年，期刊，卷，页） Cited references（Author，Year，Journal，Volume，Page）	聚类 # Cluster #
1	64.88	Arroyo M，2003，Polymer，V44，P2447	3
2	46.44	Turjanmaa K，1987，Contact Dermatitis，V17，P270	0
3	41.94	Lagier F，1992，Journal of Allergy and Clinical Immunology，V90，P319	0
4	38.28	Czuppon A B，1993，Journal of Allergy and Clinical Immunology，V92，P690	1
5	37.20	Blanco C，1994，Annals of Allergy，V73，P309	1
6	37.04	Varghese S，2003，Polymer，V44，P4921	3
7	36.82	Ziegler A D，2009，Science，V324，P1024	15
8	35.88	Potts J R，2012，Macromolecules，V45，P6045	2
9	34.83	Ray S S，2003，Progress in Polymer Science，V28，P1539	3

7.3.4　网络中高Sigma值文献

Sigma指标同时测度了引文的中心性和突发性两个属性（Chen，2012a，2017；Chen et al.，2012b；Chen and Leydesdorff，2014）。如果一篇引文在两个属性中的数值都高，而不仅是在其中一个属性中数值高，那么它将具有更高的Sigma值（表7–5）。Ko等（2003）在《Plant Molecular Biology》发表的开创性论文Sigma值最高，为518.91，报道了通过胶乳转录组分析揭示橡胶转移酶是橡胶粒子上的顺式异戊烯基转移酶。Sigma值第2的是van Beilen等（2007）在《Trends in Biotechnology》上发表的一篇综述，报道了利用新的产胶植物作为合成天然橡胶的替代来源。这意味着这些文献在结构上是必不可少的，并且其在强大的引文突发方面具有启发性。

表7-5 1988—2017年天然橡胶领域结构和时间上具有显著性的引文

Tab. 7-5 Structurally and temporally significant references of natural rubber domain during 1988-2017

排名 Rank	Sigma值 Sigma	突现值 Citation bursts	中心度 Centrality	引用次数 Citation counts	被引文献（作者，年，期刊，卷，页） Cited references（Author，Year，Journal，Volume，Page）	聚类# Cluster #
1	518.91	25.31	0.28	59	Ko J H，2003，Plant Molecular Biology，V53，P479	8
2	106.26	20.11	0.26	58	van Beilen J，2007，Trends in Biotechnology，V25，P522	8
3	11.81	12.47	0.22	28	Bokobza L，2008，Journal of Polymer Science Partb-Polymer Physics，V46，P1939	2
4	11.66	9.96	0.28	27	Tarachiwin L，2005，Biomacromolecules，V6，P1858	8

7.4 天然橡胶文献时间线网络分析

时间线可视化图谱表示各聚类沿水平时间线的分布（图7-2）。每个聚类从左到右显示，发表时间图例显示在视图顶部。聚类从0开始编号，即#0是最大的聚类，#1是第二大聚类，各聚类按其大小降序垂直排列，最大的聚类显示在视图最上方。彩色曲线代表在相应颜色年份添加的共引连线。具有红色年轮的大型结点特别值得注意，因为它们要么被高度引用，要么有引文突发，或者二者兼具（Chen，2012a，2017；Chen et al.，2012b；Chen and Leydesdorff，2014）。在每条时间线下方，列出了每一年中被引次数最高的3篇文献。在同一年的引文中，引用次数最高的文献放在最底位置，引用次数较低的文献被移到左边。

如时间线视图7-2所示，各聚类的可持续性是不同的，因此各聚类的持续时间也值得关注（表7-6）。有些聚类的周期超过20年，而有些聚类的周期相对较短；一些聚类一直活跃到2016年，是本章中引用文献最新的一年。最大的8个聚类成员均超过50个，而最大聚类的平均轮廓值略低于其他聚类。最大的聚类#0持续22年，但已经不再活跃。聚类#2跨越了18年的时间，而且仍然是活跃的。相比之下，聚类#7（天然橡胶生物合成）在2006年就结束了，但由此看到，相关研究在新的领域中找到了方向，转向橡胶延伸因子（聚类#8）研究中。

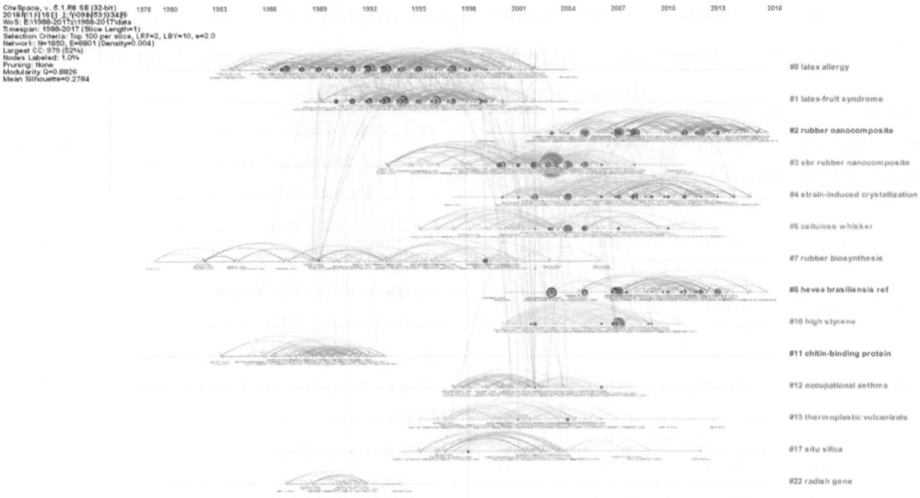

图7-2 1988—2017年天然橡胶文献时间线网络图谱

Fig. 7-2 Timeline visualization view of natural rubber literature during 1988-2017

本章主要对前12个聚类进行分析。一个研究课题或一个研究领域可以被它的知识基础和研究前沿所表征。知识基础是相关研究团体所引用的学术著作的集合，而研究前沿是受知识基础影响的延伸。不同的研究前沿可能从一个共同的知识基础发展起来（Chen，2012a，2017；Chen and Leydesdorff，2014）。

表7-6 各聚类的时间属性

Tab. 7-6 Temporal properties of major clusters

聚类 # Cluster #	大小 Size	轮廓值 Silhouette	时间跨度 From to	持续时间 Duration	平均（年） Median	可持续性 Sustainability	活跃度 Activeness	主题 Theme
0	196	0.788	1983—2004	22	1993	++++++	不活跃 Inactive	latex allergy 胶乳过敏
1	115	0.831	1988—2004	17	1995	++	不活跃 Inactive	latex-fruit syndrome 乳胶—水果综合征
2	107	0.925	1999—2016	18	2010	+++	活跃 Active	rubber nanocomposite 天然橡胶纳米复合材料

（续表）

聚类 # Cluster #	大小 Size	轮廓值 Silhouette	时间跨度 From to	持续时间 Duration	平均(年) Median	可持续性 Sustainability	活跃度 Activeness	主题 Theme
3	97	0.882	1993—2011	19	2002	++++	不活跃 Inactive	sbr rubber nanocomposite 丁苯橡胶纳米复合材料
4	74	0.938	1997—2015	19	2006	++++	活跃 Active	strain-induced crystallization 应变诱导结晶
6	54	0.972	1995—2013	19	2004	++++	不活跃 Inactive	cellulose whisker 微晶纤维素
7	54	0.952	1979—2006	28	1993	+++++++++++	不活跃 Inactive	rubber biosynthesis 橡胶生物合成
8	53	0.982	2000—2016	17	2009	++	活跃 Active	*hevea brasiliensis* ref 橡胶延伸因子
10	48	0.97	2000—2011	12	2005		不活跃 Inactive	high styrene rubber 高苯乙烯橡胶
11	46	0.991	1983—1993	11	1989		不活跃 Inactive	chitin-binding protein 几丁质结合蛋白
12	46	0.97	1997—2008	12	2002		不活跃 Inactive	occupational asthma 职业性哮喘
13	45	0.989	1995—2013	19	2002	++++	不活跃 Inactive	thermoplastic vulcanizate 热塑性硫化橡胶
17	30	0.993	1994—2012	19	2001	++++	不活跃 Inactive	situ silica 原位二氧化硅
22	14	0.999	1987—1994	8	1990		不活跃 Inactive	radish gene 萝卜基因

7.4.1 聚类#0——天然橡胶胶乳过敏

聚类#0是最大的聚类，时间跨度从1983—2004年，在22年期间包含196篇引文。该聚类的轮廓值为0.788，是主要聚类中最低的，但这通常被认为有相对较高的同质性。该聚类已经不再活跃，但在1987—1998年12年的时间是一个相当活跃的时期（图7-3），主题层次结构有两个分支：IgE介导的胶胶乳过敏和胶乳过敏的潜在风险因素。

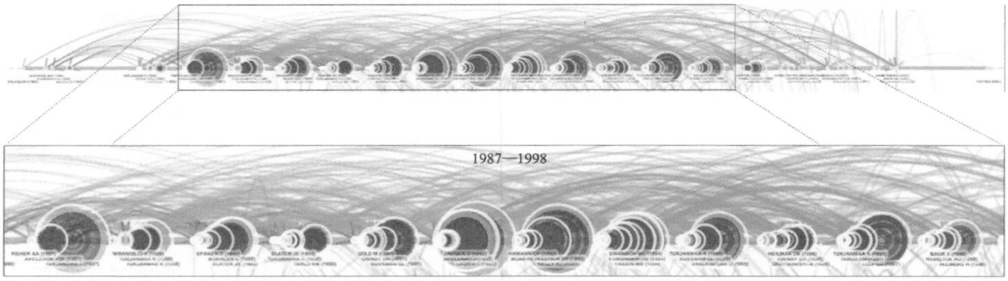

图7-3　聚类#0天然橡胶胶乳过敏高影响力引文

Fig. 7-3　High-impact members of cluster #0

• IgE介导的胶乳过敏：采用ELISA测定特定胶乳抗原和血清方法对脊柱裂患者、卫生保健工作者和其他有胶乳过敏症状人群的两种胶乳过敏原进行检测（Kelly et al.，1993）；检测乳胶手套中的过敏原与皮肤接触导致胶乳过敏的发生（Sussman et al.，1991），通过对医院员工调查，并对6种常见的空气过敏原、1种非乳胶合成手套和4种不同乳胶手套提取物进行皮肤点刺试验，研究IgE介导的胶乳过敏的发生率（Yassin et al.，1994）；比较一次性乳胶手套和其他橡胶医疗器材提取物中胶乳过敏原和总蛋白水平，发现低致敏性手套可能含有大量的IgE结合蛋白（Yunginger et al.，1994）。

• 胶乳过敏的潜在风险因素：评估手术室医护人员、放射科和外科医生胶乳过敏的发生风险（Arellano et al.，1992；Lagier et al.，1996），通过对受试者进行定期的皮肤点刺试验，揭示长期频繁接触乳胶制品（如乳胶手套、外科手术或导尿管等），以及脊柱裂患者导致胶乳过敏的潜在风险因素：超过5种手术，特异反应倾向性，历史临床症状，≥4种致敏反应（Moneret-Vautrin，1993；Michael et al.，1996）。

7.4.2　聚类#1——乳胶—水果综合征

聚类#1是第二大聚类，包含115篇引文，范围从1988—2004年共17年。该聚类的轮廓值为0.831，有较高的同质性。与聚类#0相似，是早期的专业领域，在1991—2000年10年的时间有一些高影响力的论文出现（图7-4），总体主题层次有两个：胶乳过敏原的识别和胶乳过敏原的交叉反应。

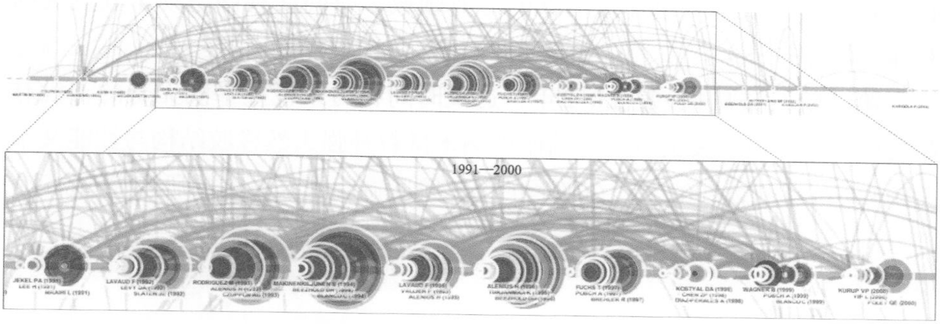

图7–4 聚类#1乳胶—水果综合征高影响力引文

Fig. 7–4 High-impact members of cluster #1

• 胶乳过敏原的识别：从天然胶乳和乳胶手套中分离出一种主要过敏原蛋白，经鉴定是橡胶延伸因子，命名为Hev b1（Czuppon et al.，1993）；鉴定了天然胶乳中3种主要过敏原：橡胶树凝集因子、β-1，3-葡聚糖酶和几丁质酶，分别命名Hev b2、Hev b6和Hev b11（Alenius et al.，1995）；克隆了天然胶乳中小橡胶粒子蛋白上的主要致病原Hev b3基因，构建其原核表达载体并表达和纯化重组蛋白（Wagner et al.，1999）；分离纯化了天然胶乳中的过敏原Hev b8，是一种胶乳抑制蛋白，也存在于香蕉提取物中，可能导致交叉反应（Vallier et al.，1995）；克隆了存在于天然胶乳和乳胶制品中的主要过敏原Hev b5基因，与猕猴桃中的一种酸性蛋白有高度同源性（Akasawa et al.，1996）。

• 胶乳过敏原的交叉反应：发现胶乳过敏患者对一些水果产生交叉过敏反应，如对鳄梨、栗子、香蕉、猕猴桃和木瓜等也发生过敏反应，导致乳胶—水果综合征（Blanco et al.，1994；Beezhold et al.，1996）；通过抑制试验证明胶乳交叉反应的IgE抗体能识别胶乳和水果（木瓜、鳄梨、香蕉、栗子、百香果、无花果、芒果、猕猴桃、菠萝、桃、番茄等）中的过敏原（Brehler et al.，1997）。

7.4.3 聚类#2——天然橡胶纳米复合材料

聚类#2是第三大聚类，有107篇引文，其轮廓值为0.925，比前两个大的聚类#0和聚类#1高，这说明该聚类的同质性更高。该聚类从1999—2016年活跃了18年，代表了一个非常活跃的专业领域，时间线视图呈现了其发展的3个阶段（图7-5）。

第一阶段是从1999—2002年，这段时间相对来说是平淡无奇的，没有高的引用次数或引文突发。一篇刚性纳米微粒补强天然橡胶结构与性能文章（Bokobza et al.，2002），在随后的第二阶段掀起了高影响的研究浪潮。

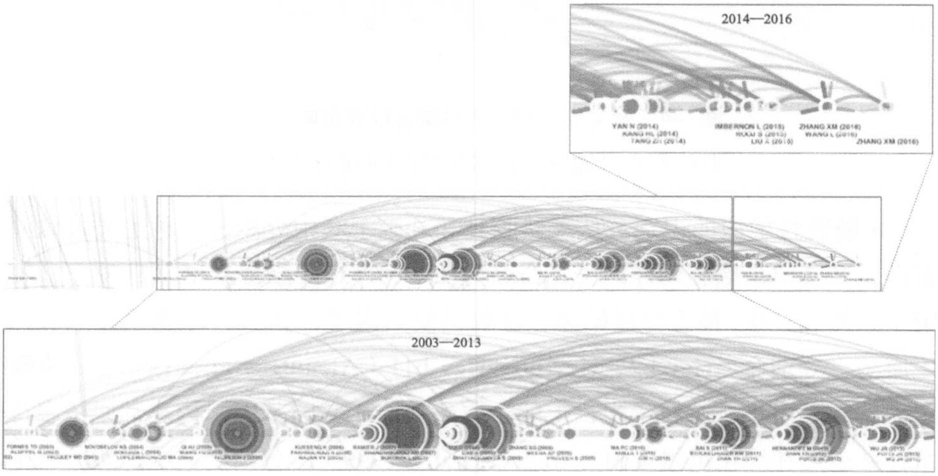

图7-5　聚类#2天然橡胶纳米复合材料高影响力引文

Fig. 7-5　High-impact members of cluster #2

第二阶段是2003—2013年。与第一阶段不同的是，第二阶段充满了高影响力的贡献——大量的引文年轮和引文突发以红色显示，在这一时期出现了几类高影响力的论文：

• 层状黏土/天然橡胶纳米复合材料：蒙脱土（Fornes et al.，2003）、累托石（Wang et al.，2005）、炭黑/纳米黏土/丁苯橡胶（Praveen et al.，2009）、层状硅酸盐/丁苯橡胶（Zhang et al.，2009）、纳米SiO_2（Meera et al.，2009）等作为填料制备天然橡胶纳米复合材料。

• 碳纳米管/天然橡胶纳米胶复合材料：单壁碳纳米管（Frogley et al.，2003；López-Manchado et al.，2004；Kueseng et al.，2006）、单壁碳纳米管与SIC碳纳米颗粒（Kueseng et al.，2006），多壁碳纳米管（Bokobza，

2007；Fakhru'l-Razi et al.，2006；Shanmugharaj et al.，2007）、改性碳纳米管（Bhattacharyya et al.，2008；Das et al.，2008）充填天然橡胶制备纳米复合材料及其机械性能、导热性、光谱性研究。

• 石墨烯/天然橡胶纳米复合材料：石墨烯（Kim et al.，2010；Wu et al.，2013）、氧化石墨烯（Wu et al.，2013；Potts et al.，2013）补强天然橡胶纳米复合材料及其物理机械性能、电学性能、气体阻隔性能、热学性能研究，相应制备方法有机械共混（Kim et al.，2010）、乳液共混（Zhan et al.，2011）、溶液共混（Bai et al.，2011）和原位聚合（Zhan et al.，2012）。

第三阶段是2014—2016年，到目前为止已有一些引文突发，这一时期的研究主题进一步揭示了该领域最近的发展状况。在这一时期，大多数引文涉及石墨烯或氧化石墨烯的加工和化学表征，作为填料制备高阻隔、高机械性能和高电热屏蔽性的改性石墨烯或改性氧化石墨烯/天然橡胶纳米复合材料及填料网络的形成对提高材料整体性能的作用（Tang et al.，2014；Zhang et al.，2016；Imbernon and Norvez，2016）。

7.4.4　聚类#3——丁苯橡胶/天然橡胶纳米复合材料

聚类#3是第四大聚类，包含97篇引文，其持续时间范围从1993—2011年（图7-6）。2000—2008年的9年时间是一个非常活跃的时期，在这一时期最突出的贡献包括高被引文献和高覆盖率的施引文献。首先介绍了以丁二烯橡胶和丁苯橡胶为基础，制备有机层状硅酸盐橡胶复合材料（Ganter et al.，2001）；随后开展了以纳米材料层状硅酸盐为填料增强丁苯橡胶/天然橡胶复合材料研究，通过有机改性蒙脱土替代炭黑与天然橡胶插层复合，片层以纳米尺寸分散于橡胶之中，形成一种有良好力学、热稳定、气体阻隔及加工性能的纳米复合材料（Joly et al.，2002；Arroyo et al.，2003）；采用胶乳凝聚法制备二氧化硅/天然橡胶纳米复合材料（Bokobza et al.，2005）；硅烷偶联剂有机改性膨润土填充丁苯橡胶制备纳米复合材料及其加工、力学和阻隔性能研究，数学模型和响应面法评估有机黏土/丁苯橡胶纳米复合材料的拉伸性、焦灼时间和门尼黏度（Chakraborty et al.，2010a；2010b）。

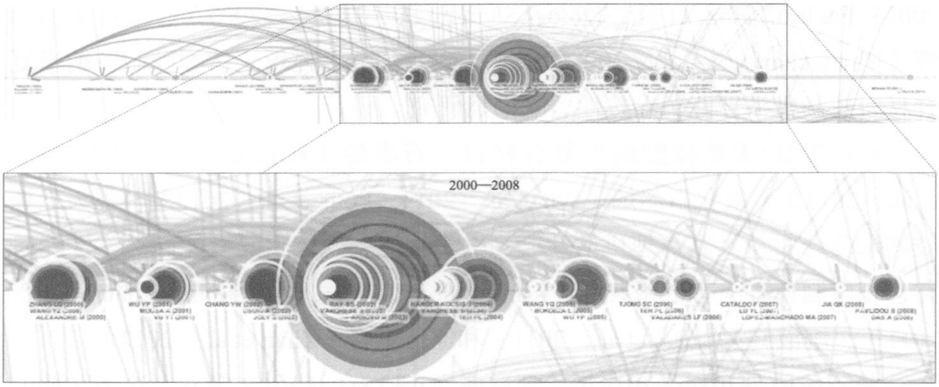

图7-6　聚类#3丁苯橡胶/天然橡胶纳米复合材料高影响力引文

Fig. 7-6　High-impact members of cluster #3

7.4.5　聚类#4——应变诱导结晶

聚类#4是第五大聚类，包含74篇引文，持续时间为1997—2015年，是一个仍在活跃的聚类，时间线视图将其分为3个阶段（图7-7）。

第一阶段是1997—2001年，出现两篇突出的引用，一篇是天然橡胶拉伸取向和应变诱导结晶试验（Toki et al.，2000），另一篇是发现饱和脂肪酸与加速天然橡胶分子链的诱导结晶有关（Tanaka，2001）。

在第一阶段研究基础上，第二阶段2002—2011年出现如下两类研究：

• 未增强天然橡胶拉伸取向和应变诱导结晶：通过原位同步电子衍射测定在单轴变形过程中硫化天然橡胶的分子取向和应变诱导结晶，发现这些优异性能来源于其在拉伸过程中的结晶能力（Toki et al.，2002）；同步辐射二维广角X射线衍射分析了未硫化和硫化天然橡胶的应变诱导结晶行为（Tosaka et al.，2004）；对Mullins效应的恢复研究表明，填充橡胶和天然橡胶的强度在室温下仅能部分恢复，在高温、真空或与溶剂接触条件下可显著加速Mullins效应的恢复（Diani et al.，2009）；对天然橡胶的结晶结构、应变诱导结晶与张力—收缩试验关系，特别是由应变诱导结晶作用产生的应力—应变曲线滞后现象进行研究（Huneau，2011）。

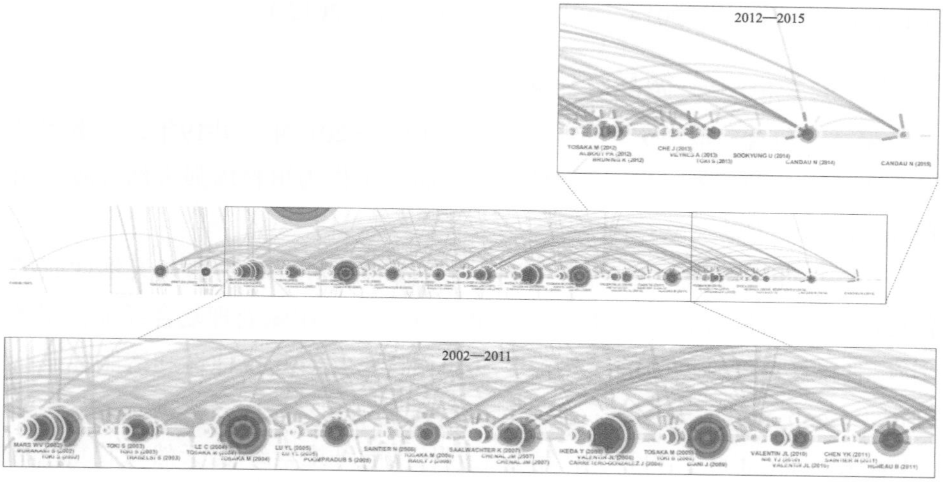

图7-7 聚类#4应变诱导结晶高影响力引文

Fig. 7-7 High-impact members of cluster #4

• 增强天然橡胶拉伸取向和应变诱导结晶：一是炭黑增强天然橡胶，研究炭黑增强天然橡胶在22℃的拉伸结晶，发现天然橡胶在应变为4h开始出现结晶，而添加炭黑后在应变约为2h开始出现结晶，炭黑的加入使晶体尺寸及晶体取向都变小，影响结晶取向（Trabelsi et al.，2003）；炭黑和碳酸钙增强天然橡胶，发现橡胶拉伸结晶对其拉伸强度有很大影响（Poompradub et al.，2005）；在弱、高硫化炭黑、二氧化硅和接枝硅填充天然橡胶的拉伸过程中，进行原位同步电子衍射试验，在一个拉伸周期后观察到Mullins效应改变了样品的应变诱导结晶行为（Chenal et al.，2007）。二是纳米黏土增强橡胶，在交联天然橡胶中加入黏土纳米粒子，提供了一种更均匀地分布式网络结构，并能在单轴变形下提高拉伸取向和应变结晶（Carretero-González et al.，2008）。

2012—2015年代表该专业领域最新的发展动向。采用原位同步X射线衍射研究不同温度下，未硫化、硫化天然橡胶、异戊橡胶的应力与应变关系及应变诱导结晶；在室温和低应变频率下，采用原位广角X射线衍射研究天然橡胶的应变诱导结晶特点和周期性变形；开发一种频闪X射线衍射仪研究天然橡胶应变诱导结晶的动力学特性；研究应变速率（$5.6 \times 10^{-5} \sim 2.8 \times 10^{1}\ s^{-1}$）和温度（−40 ~ 80℃）对应变诱导天然橡胶结晶发生的影响（Toki et al.，

2013；Candau et al.，2014，2015；Albouy et al.，2012）。

7.4.6 聚类#6——微晶纤维素

聚类#6包含54篇引文，其持续时间从1995—2013年，共19年，该聚类已经不再活跃（图7-8）。主题领域是天然高分子作为填料增强天然橡胶，出现几类高影响力的文献：

• 植物纤维补强天然橡胶：研究硅烷偶联剂对竹纤维填充天然橡胶硫化特性和力学性能的影响（Ismail et al.，2002）；用碱处理复合纤维，制备剑麻/油棕复合纤维增强的天然橡胶复合材料，并对其拉伸性和溶胀性进行分析（Jacob et al.，2004）；研究椰壳纤维补强天然橡胶及其动态力学性能（Geethamma et al.，2005）；制备不同长度的纤维素纤维/天然橡胶复合材料，并比较其在机械性能、热性能、吸水性和热塑性的差异（Abdelmouleh et al.，2007）。

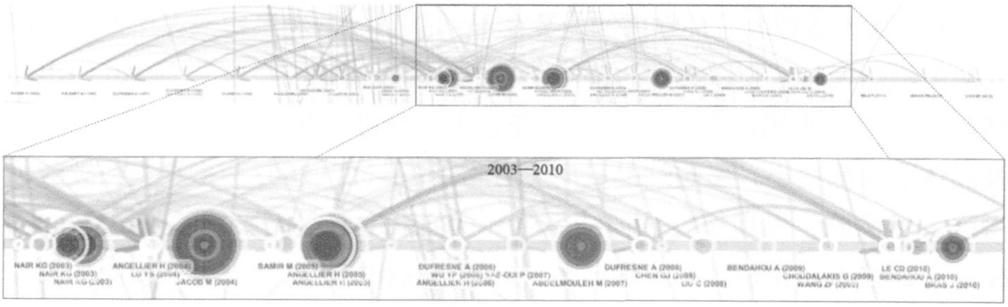

图7-8 聚类#6微晶纤维素高影响力引文

Fig. 7-8 High-impact members of cluster #6

• 纳米微晶纤维素补强天然橡胶：从蟹壳中提取纳米微晶纤维素，并将其用作生物基天然橡胶的补强填料，通过乳液共混法制备天然橡胶/纳米微晶纤维素复合材料（Nair et al.，2003）；采用硫酸水解淀粉纳米微晶纤维素补强天然橡胶基复合材料，并对其水汽和氧的阻隔性进行研究（Angellier et al.，2005）；采用酸水解甘蔗渣制得微晶纤维素补强天然橡胶，制备纳米复合薄膜，发现纤维素的长径比对材料的拉伸性能影响较大，并提高了材料的机械性能和土壤中的降解速率（Bras et al.，2010）。

• 淀粉补强天然橡胶：通过化学改性淀粉与聚丙烯酸丁酯接枝作为填料增

强天然橡胶及其抗拉伸、抗撕裂和抗断裂性能的研究（Liu et al.，2008）。

7.4.7 聚类#7——天然橡胶生物合成

聚类#7包含54篇引文，时间范围从1979—2006年持续28年，是各聚类中持续时间最长的，但该聚类或其专业领域在很大程度上是不活跃的（图7-9）。该聚类时间线视图中有两个突出的引用，一篇是证明橡胶延伸因子（REF）是一种与橡胶粒子紧密结合的橡胶粒子蛋白，主要作用是参与橡胶生物合成中分子链的延伸过程（Dennis et al.，1989）；另一篇是发现小橡胶粒子蛋白（SRPP）可以促进天然橡胶的合成（Oh et al.，1999）。引用了该聚类的施引文献，也在相当程度上揭示了与主题的一致性。认为法尼基焦磷酸合成酶存在于C-乳清中，它催化牻牛儿基焦磷酸形成法尼基焦磷酸，使异戊烯基焦磷酸（IPP）掺入到聚异戊二烯（Light et al.，1989）；发现含顺式橡胶的植物体内有顺式-异戊烯基转移酶，紧密地结合在胶粒的膜嵌合体上，催化顺式-1，4-异戊二烯聚合到延伸中的天然橡胶分子链上（Cornish，1993；2001）；对不同产胶植物，如橡胶树、银胶菊、无花果树、印度榕天然橡胶合成的分子量进行调控（Kang et al.，2000a；2000b）。

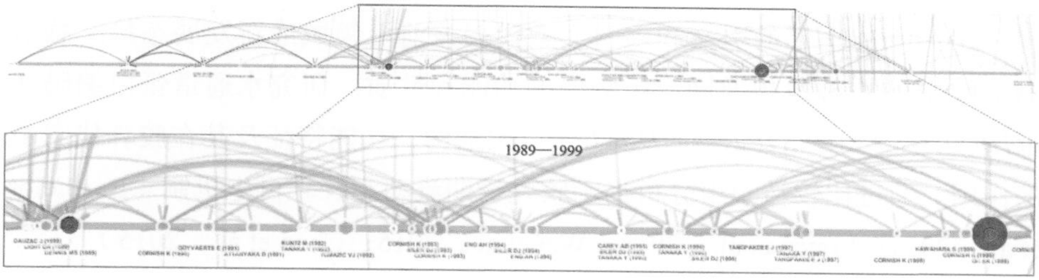

图7-9 聚类#7天然橡胶生物合成高影响力引文

Fig. 7-9 High-impact members of cluster #7

7.4.8 聚类#8——橡胶延伸因子

聚类#8包含53篇引文，时间范围从2000—2016年活跃了17年，是一个仍在持续发展的聚类（图7-10）。在2003—2013年11年的时间里出现结构和时间上具有重要性的文献，这些文献不仅将不同聚类连接起来，起到桥梁和枢纽作用，还是#7在新的研究领域的延续，它们有着共同的知识基础。从胶乳cDNA文库中获得了4个顺式-异戊烯基转移酶的cDNA序列，分别命名为*HbCPTl*、*HbCPT2*、*HbCPT3*和*HbCPT4*，发现*HbCPT3*和*HbCPT4*的酶活性具有把^{14}C-IPP掺入到法尼基焦磷酸（FPP）中的能力（Ko et al.，2003）；经磷脂酶处理的天然胶乳的分子量和哈金斯参数都会减少，可能是由于连接在天然橡胶α末端的磷酸单酯通过胶束形式或磷脂分子的极性端形成支点，增大了天然橡胶的分子量（Tarachiwin et al.，2005）；发现SRPP对橡胶生物合成效率和合成稳定性具有非常关键的作用（Hillebrand et al.，2012）。

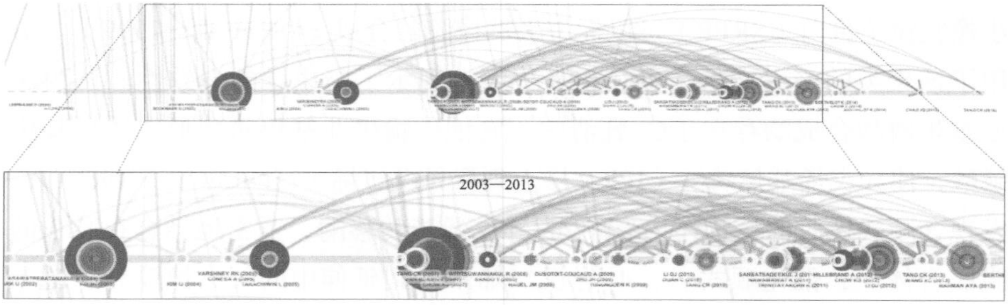

图7-10　聚类#8橡胶延伸因子高影响力引文

Fig. 7-10　High-impact members of cluster #8

2014—2016年的引文和施引文献代表该前沿领域最新发展态势。总结REF和SRPP在橡胶树及同源植物中的功能和性质。研究水通道蛋白编码基因*HbXIP1; 1*、*HbXIP2; 1*和*HbXIP3; 1*的结构、功能特征及其在胶乳代谢中的作用。通过比较转录组分析揭示胶乳长期流动和胶乳再生机制（Berthelot et al.，2014；Lopez et al.，2016；Wei et al.，2015；Chao et al.，2015）。

7.4.9　聚类#10——高苯乙烯橡胶

聚类#10包含48篇引文，时间范围从2000—2011年持续了12年，是一个不再活跃的聚类（图7-11）。但在2002—2010年间出现了高影响力的引文，这些文献是天然橡胶在高分子材料科学研究领域的补充，在当时被大量引用。

Leblanc讨论橡胶与填料相互作用的性质及其对未固化材料流变性能的影响。描述了填充橡胶聚合物所表现出的流变性能，然后从填料特性的角度综述了填料与弹性体之间的相互作用，对填充橡胶的尺寸方面进行了详细的讨论，为理解黏合橡胶和流动性能之间的关系提供了关键证据（Leblanc,

2002）。Choi等研究了填料—填料相互作用对炭黑和二氧化硅填充天然橡胶流变性能的影响。发现炭黑/二氧化硅比例为40/40和20/60phr的填充复合物表现出异常的流变行为，在测量过程中门尼黏度突然升高，然后在某一时刻又下降。这种异常行为可以用强填料—填料相互作用来解释（Choi et al.，2003）。Arroyo等分析环氧化天然橡胶（ENR）和填料处理对天然橡胶（NR）纳米复合材料形态和行为的影响，制备了几种聚合物共混物。用X射线衍射法测定了填充后黏土的分散性和分散度。通过流变学、力学和溶胀特性分析了黏土在橡胶中的作用。结果表明，硅酸盐纳米层在橡胶基体中的分散程度、有机黏土类型和弹性体相容性是影响其性能的重要因素（Arroyo et al.，2007）。Rattanasom等采用二氧化硅/炭黑复合填料在不同配比下对天然橡胶进行增强，以确定最佳的二氧化硅/炭黑比例。并测定了天然橡胶硫化胶的抗拉强度、撕裂强度、耐磨性、抗裂纹扩展阻力、热积累阻力、抗滚动阻力等力学性能（Rattanasom et al.，2007）。Rajasekar等以环氧化天然橡胶为相容剂，以硫为固化剂，制备环氧化天然橡胶/有机改性纳米黏土复合材料（EC）。形态研究表明，ENR基质中纳米黏土插层以及丁腈橡胶（NBR）基质中EC的进一步掺入导致纳米黏土的脱落。固化研究表明，与纯丁腈橡胶相比，加入EC的丁腈橡胶具有更快的焦化时间、固化时间和最大转矩。动态力学分析表明，丁腈橡胶基体中含有EC载荷的化合物的贮存模量增加，阻尼特性减弱（Rajasekar et al.，2009）。

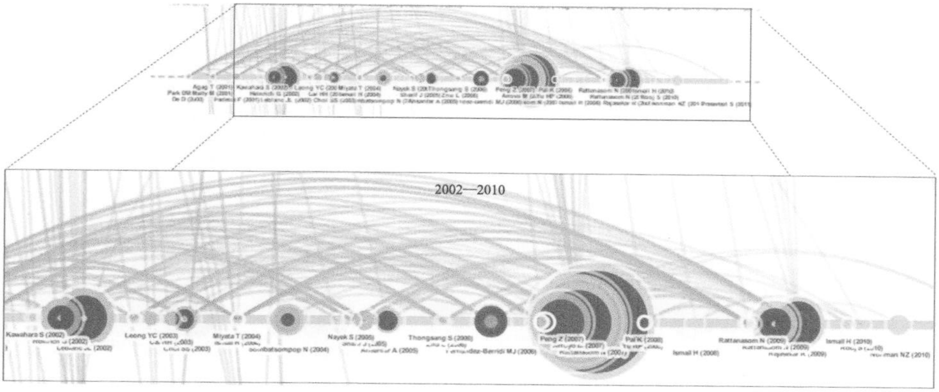

图7-11　聚类#10高苯乙烯橡胶高影响力引文

Fig. 7-11　High-impact members of cluster #10

7.4.10 聚类#12——职业性哮喘

聚类#12包含46篇引文，从1997—2008年持续了12年，涉及早期的天然胶乳过敏研究领域，在2001—2007年出现引用激增（图7-12）。Allmers等评估干预措施的效果，如对医生和管理人员进行教育，要求只购买低蛋白、无粉的NRL手套以减少医疗保健工作人员胶乳过敏的发病率。发现疑似职业性天然胶乳NRL过敏病例的发生率在1998年前一直在上升，此后一直在稳步下降。从购买粉末NRL检查手套数量开始下降到怀疑由NRL引起的职业性哮喘病例数量开始下降之间有两年的滞后期（Allmers et al.，2002）。Vandenplas等比较胶乳诱发哮喘患者减少或停止接触乳胶制品前后的健康状况，减少暴露于乳胶制品是一个安全的选择（Vandenplas et al.，2002）。Sastre为了确定一种标准化的胶乳提取物特异性免疫疗法在致敏工人中的有效性和安全性。对NRL过敏的接触性荨麻疹患者、鼻炎或哮喘患者分组治疗。临床疗效主要表现在皮肤症状上，而鼻炎和哮喘症状在特定的吸入治疗中也有改善。乳胶特异性免疫治疗是治疗致敏工人乳胶过敏的一种有效方法（Sastre et al.，2003）。Bousquet等研究与工作相关的乳胶接触对胶乳过敏产生的风险进行评估。在卫生保健工作者和普通人群中比较乳胶过敏的患病率和发病率，对卫生保健工作者的暴露—反应关系进行评估（Bousquet et al.，2006）。

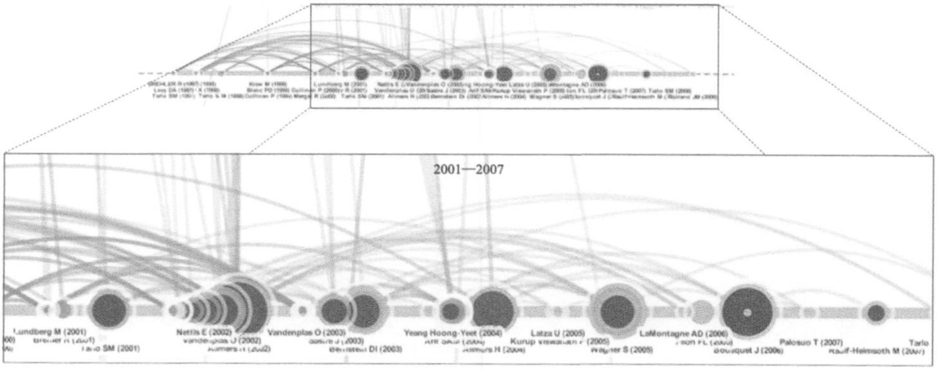

图7-12　聚类#12职业性哮喘高影响力引文

Fig. 7-12　High-impact members of cluster #12

7.4.11 聚类#13——热塑性硫化橡胶

聚类#13包含43篇引文，从1995—2013年持续了19年，是天然橡胶复合材料领域的延伸，在1996—2006年出现引用激增（图7-13）。Nakason等在甲苯溶液中合成天然橡胶与马来酸酐（NR-g-MA）或马来酸化天然橡胶（MNR）的共聚物。采用过氧化苯甲酰（BPO）引发自由基接枝共聚。考察单体浓度、引发剂浓度、反应温度和反应时间对反应的影响。以75/25 ENR/PP共混物为原料，以Ph-PP为增容剂，合成了摩尔比为30的环氧化天然橡胶，并用其制备了热塑性硫化胶。对不同固化体系（即硫、过氧化物及硫和过氧化物混合固化体系）的影响进行了研究。发现混合固化体系的混合转矩、剪切应力、剪切黏度、拉伸强度和断裂伸长率分别高于硫和过氧化物固化体系。这可能是于在ENR阶段形成了S-S、C-S与C-C键。与混合硫化体系相比，过氧硫化体系的热塑性硫化橡胶（TPV）的分散硫化橡胶颗粒范畴更小（Nakason et al.，2004；2006）。Thiraphattaraphun等以过硫酸钾为引发剂，采用乳液聚合法将甲基丙烯酸甲酯单体接枝到天然橡胶上。考察了引发剂浓度、反应温度、单体浓度和反应时间对单体转化率和接枝效率的影响，采用溶剂萃取法测定了接枝效率。采用熔融混合法制备了天然橡胶-g-甲基丙烯酸甲酯/聚甲基丙烯酸甲酯（NR-g-MMA/PMMA）共混体系。研究了接枝共聚物组成和共混比对GNR/PMMA共混物的力学性能和断裂行为的影响（Thiraphattaraphun et al.，2001）。Kawahara等以叔丁基氢过氧化物/四乙烯五胺为引发剂，研究了脱蛋白天然胶乳与苯乙烯接枝共聚制备纳米基质分散聚合物。发现天然橡胶以颗粒直径0.5μm，分散在厚度约15nm的聚苯乙烯基质中（Kawahara et al.，2003）。Sirisinha等测定了天然橡胶部分取代弹性体氯化聚乙烯（CPE）的性能，包括共混物的力学性能、热老化性能和耐油性能，发现共混物中NR的含量对共混物的性能影响较大。还研究了EPDM-g-MA作为相容剂和酚类抗氧化剂对50/50 CPE/NR共混物热稳定性的影响，结果表明，EPDM-g-MA可使共混体系的相尺寸减小，说明其具有增容作用。此外，抗油和热老化性能的结果与形态学结果吻合较好，说明在50/50 CPE/NR共混体系中，相对抗拉强度对抗油和热老化性能的影响强烈受CPE基体中NR分散相的大小控制，分散相尺寸越小，抗油和热老化的能力越强（Sirisinha et al.，2002；2004）。

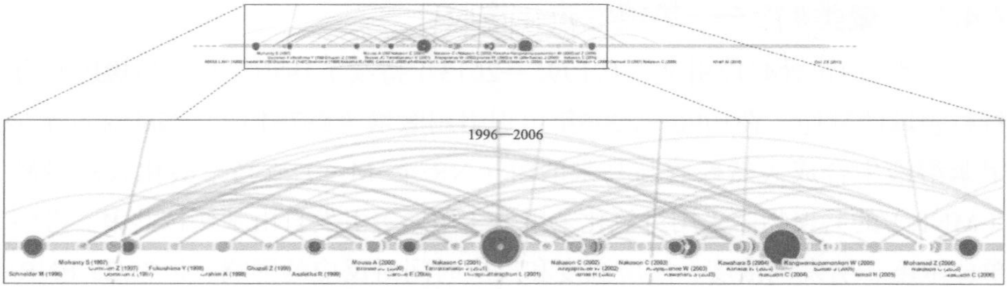

图7-13　聚类#13热塑性硫化橡胶高影响力引文

Fig. 7-13　High-impact members of cluster #13

7.4.12　聚类#17——原位二氧化硅

聚类#17包含30篇引文，从1994—2012年持续了19年，是天然橡胶复合材料领域的一个组成部分，在1998—2004年出现引用激增（图7-14）。Hashim等研究发现，四硫醚（TESPT）对苯乙烯—丁二烯橡胶（SBR）和原位填充二氧化硅基的SBR硫化胶的固化特性和物理性能有影响（Hashim et al.，1998）。Kohjiya等综述了在没有硅烷偶联剂的情况下，溶胶—凝胶法在通用级橡胶上的应用。四乙氧基硅烷（TEOS）在环氧化天然橡胶（ENR）、苯乙烯—丁二烯橡胶（SBR）或丁二烯橡胶（BR）中的溶胶—凝胶反应生成原位二氧化硅。发现原位生成的二氧化硅是一种良好的增强剂，在橡胶基体中的分散性变好，粒径变小，呈单分散性，提高了材料的力学性能。通过四乙氧基硅烷的溶胶—凝胶反应，在天然橡胶（NR）“绿色”基体中产生了细小、分散良好的原位二氧化硅颗粒，并对天然橡胶/二氧化硅复合材料进行了研究，发现原位二氧化硅对天然橡胶硫化胶具有良好的增强效果（Kohjiya and Ikeda，2000；Kohjiya et al.，2001）。Park等研究硅烷偶联剂处理的二氧化硅/橡胶复合材料的交联密度和热稳定性。结果表明，硅烷表面处理使二氧化硅表面的有机官能团增加了二氧化硅与橡胶基体之间的界面附着力（Park and Cho，2003）。Ikeda等以天然橡胶（NR）为基体，采用四乙氧基硅烷（TEOS）进行溶胶—凝胶反应，制备了不同含量的原位二氧化硅纳米粒子/天然橡胶“绿色”复合材料，而原位二氧化硅对NR硫化胶的增强效果随原位二氧化硅含量的增加而增大（Ikeda and Kameda，2004）。

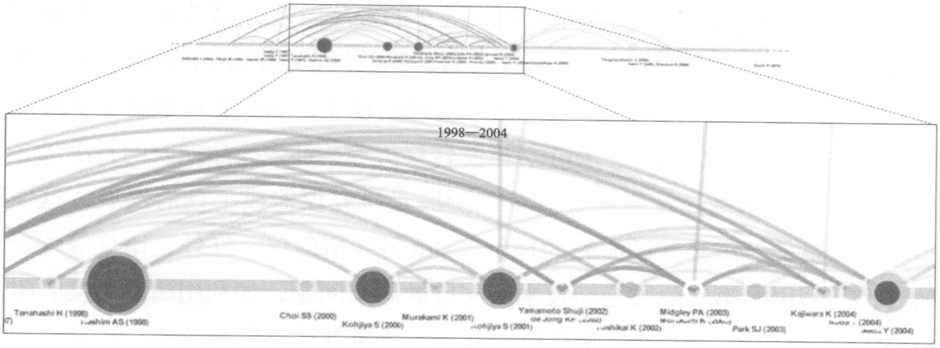

图7-14 聚类#17原位二氧化硅高影响力引文

Fig. 7-14 High-impact members of cluster #17

7.4.13 其余的聚类

其余的聚类要么规模相对较小，要么持续时间较短。本章将省略对这些聚类的详细讨论。聚类#11关注几丁质酶结合蛋白，该聚类包含46篇引文，从1983—1993年持续了11年，是早期的研究领域。该专业领域的主要贡献者，如Chrispeels和Raikhel综述了凝集素、凝集素基因及其在植物防御中的作用（Chrispeels and Raikhel，1991），对橡胶树在生物科学领域的开创研究作出积极贡献。Broglie、Neuhaus等关于受乙烯调节的几丁质酶基因的表达，几丁质酶在植物液泡中的靶向作用也是该聚类的关键成员（Broglie et al.，1986；Neuhaus et al.，1991）。聚类#23关注萝卜基因，该聚类在1987—1994年较活跃，引用的引文大多是20世纪90年代初发表的文章。

7.5 本章小结

（1）识别出天然橡胶研究的主要专业领域（研究前沿），这些专业领域在天然橡胶研究的各个方面作出积极的贡献。20世纪70年代末到90年代初，形成了以天然橡胶胶乳过敏、乳胶—水果综合征、橡胶生物合成等为主的早期前沿领域；到了20世纪90年代末至21世纪初期，形成了以丁苯橡胶或植物纤维素或微晶纤维素/天然橡胶纳米复合材料为主的中期前沿领域；进入21世纪后，形成了以纳米微粒/天然橡胶复合材料及其应变诱导结晶、橡胶延伸因子等为主的当前研究前沿。

（2）描述了天然橡胶各专业领域潜在的发展阶段，以及从一个专业领域转变到另一个专业领域的演进变化。其演变过程可分为4个阶段，第一阶段是从概念化开始，在这个阶段是研究对象的建立，如胶乳过敏的主要过敏原，参与橡胶合成的SRPP、REF、橡胶转移酶等，补强天然橡胶的石墨烯、碳纳米管、纳米微晶纤维素等。第二阶段是研究仪器或工具的发展，这使得科研人员能够解释基础研究的现象，如X射线衍射仪、红外光谱仪等仪器的开发和利用。第三阶段是新技术所支持的研究问题，如利用酶联免疫吸附ELISA法检测胶乳过敏原，采用胶乳共混、机械共混、原位改性、接枝等制备天然橡胶复合材料等。第四阶段是知识的转化，对积累的领域知识进行综合综述，并传达给现有专业成员和新的专业成员，如总结了层状硅酸盐/天然橡胶纳米复合材料的制备方法、纳米材料的准备和性质、复合材料的结晶行为、流变学特性、加工工艺等；综述了新的产胶植物银胶菊、蒲公英、印度榕等作为合成天然橡胶的替代来源等。

（3）确定了主要聚类的高影响力文献和构成研究前沿的施引文献，为天然橡胶研究提供了可靠的历史文献资料。在每个聚类中，关注的是聚类成员，这些成员是由在结构和时间上具有研究影响和演化意义的指标来测度。CiteSpace通过可视化编码设计来表示这些指标的强度，因此在这些指标中突出的文献将很容易在可视化图谱中看到（Chen，2012a，2017；Chen and Leydesdorff，2014）。一个常用的度量指标是网络中节点之间的中心性，中心性较高的节点倾向于识别出可能导致革命性变化的潜力，如Ko等（2003）和Tarachiwin等（2005）发表的论文连接了两个不同的聚类，是从橡胶生物合成转向橡胶延伸因子研究的动力。突发性探测是一种计算技术，用于识别事件和其他类型的信息的突然变化，如Dennis（1989）和Arroyo（2003）发表的论文分别在1989—1999年和2006—2013年具有强烈的引文突发，在当时掀起了研究高潮。节点的Sigma值测度的是节点之间的中心性和引文突发的组合，如Ko（2003）、Tarachiwin（2005）、van Beilen（2007）和Bokobza（2008）等发表的论文在网络结构上是必不可少的，并具有启发性意义。

（4）国内外学者对天然橡胶文献科学计量研究还鲜见报道，本章在之前对天然橡胶研究热点计量分析的基础上，进一步扩大检索范围，识别出国际天然橡胶领域的研究前沿和演进历程，对天然橡胶领域及相关学科的发展具有指导意义。引文分析可视化通过处理大量引文数据，能更容易地挖掘、识别和探测文献信息，发现数据中潜在的模式和趋势。目前应用知识图谱方法，获得了比其他方法更丰富的研究成果。

参考文献

Abdelmouleh M，Boufi S，Belgacem M N，et al.，2007. Short natural-fibre reinforced polyethylene and natural rubber composites：effect of silane coupling agents and fibres loading[J]. Composites Science and Technology，67（7）：1 627−1 639.

Akasawa A，Hsieh L S，Martin B M，et al.，1996. A novel acidic allergen，Hev b 5，in latex-Purification，cloning and characterization[J]. Journal of Biological Chemistry，271（41）：25 389−25 393.

Albouy P A，Guillier G，Petermann D，et al.，2012. A stroboscopic X-ray apparatus for the study of the kinetics of strain-induced crystallization in natural rubber[J]. Polymer，53（15）：3 313−3 324.

Alenius H，Kalkkinen N，Lukka M，et al.，1995. Prohevein from the rubber tree（*Hevea brasiliensis*）is a major latex allergen[J]. Clinical & Experimental Allergy，25（7）：659−665.

Allmers H，Schmengler J，Skudlik C，2002. Primary prevention of natural rubber latex allergy in the German health care system through education and intervention[J]. Journal of Allergy and Clinical Immunology，110（2）：318−323.

Angellier H，Molina-Boisseau S，Lebrun L，et al.，2005. Processing and structural properties of waxy maize starch nanocrystals reinforced natural rubber[J]. Macromolecules，38（9）：3 783−3 792.

Arellano R，Bradley J，Sussman G，1992. Prevalence of latex sensitization among hospital physicians occupationally exposed to latex gloves[J]. Anesthesiology，77（5）：905−908.

Arroyo M，Lopez-Manchado M A，Herrero B，2003. Organo-montmorillonite as substitute of carbon black in natural rubber compounds[J]. Polymer，44（8）：2 447−2 453.

Arroyo M，Lopez-Manchado M A，Valentin J L，et al.，2007. Morphology/behaviour relationship of nanocomposites based on natural rubber/epoxidized natural rubber blends[J]. Composites Science and Technology，67（7）：1 330−1 339.

Bai X，Wan C，Zhang Y，et al.，2011. Reinforcement of hydrogenated carboxylated nitrile-butadiene rubber with exfoliated graphene oxide[J]. Carbon，49（5）：1 608−1 613.

Beezhold D H, Sussman G L, Liss G M, et al., 1996. Latex allergy can induce clinical reactions to specific foods[J]. Clinical & Experimental Allergy, 26（4）：416–422.

Berthelot K, Lecomte S, Estevez Y, et al., 2014. *Hevea brasiliensis* REF（Hev b1）and SRPP（Hev b3）：an overview on rubber particle proteins[J]. Biochimie, 106：1–9.

Bhattacharyya S, Sinturel C, Bahloul O, et al., 2008. Improving reinforcement of natural rubber by networking of activated carbon nanotubes[J]. Carbon, I 46（7）：1 037–1 045.

Blanco C, Carrillo T, Castillo R, et al., 1994. Latex allergy-clinical-features and cross-reactivity with fruits[J]. Annals of Allergy, 73（4）：309–314.

Bokobza L, 2007. Multiwall carbon nanotube elastomeric composites：a review[J]. Polymer, 48（17）：4 907–4 920.

Bokobza L, Chauvin J P, 2005. Reinforcement of natural rubber：use of in situ generated silicas and nanofibres of sepiolite[J]. Polymer, 46（12）：4 144–4 151.

Bokobza L, Rapoport O, 2002. Reinforcement of natural rubber[J]. Journal of Applied Polymer Science, 85（11）：2 301–2 316.

Bousquet J, Flahault A, Vandenplas O, et al., 2006. Natural rubber latex allergy among health care workers：a systematic review of the evidence[J]. Journal of Allergy and Clinical Immunology, 118（2）：447–454.

Bras J, Hassan M L, Bruzesse C, et al., 2010. Mechanical, barrier, and biodegradability properties of bagasse cellulose whiskers reinforced natural rubber nanocomposites[J]. Industrial Crops and Products, 32（3）：627–633.

Brehler R, Theissen U, Mohr C, et al., 1997. "Latex-fruit syndrome"：frequency of cross-reacting IgE antibodies[J]. Allergy, 52（4）：404–410.

Broglie K E, Gaynor J J, Broglie R M, 1986. Ethylene-regulated gene expression：molecular cloning of the genes encoding an endochitinase from Phaseolus vulgaris[J]. Proceedings of the National Academy of Sciences, 83（18）：6 820–6 824.

Candau N, Laghmach R, Chazeau L, et al., 2014. Strain-induced crystallization of natural rubber and cross-link densities heterogeneities[J]. Macromolecules, 47（16）：5 815–5 824.

Candau N, Laghmach R, Chazeau L, et al., 2015. Influence of strain rate and temperature on the onset of strain induced crystallization in natural rubber[J]. European Polymer Journal, 64：244–252.

Carretero-González J, Retsos H, Verdejo R, et al., 2008. Effect of nanoclay on natural rubber microstructure[J]. Macromolecules, 41（18）：6 763–6 772.

Chakraborty S, Kar S, Dasgupta S, et al., 2010a. Effect of treatment of Bis（3-triethoxysilyl propyl）tetrasulfane on physical property of in situ sodium activated and organomodified bentonite clay–SBR rubber nanocomposite[J]. Journal of Applied Polymer Science, 116 (3): 1 660–1 670.

Chakraborty S, Kar S, Dasgupta S, et al., 2010b. Study of the properties of in-situ sodium activated and organomodified bentonite clay-SBR rubber nanocomposites-Part I: characterization and rheometric properties[J]. Polymer Testing, 29 (2): 181–187.

Chao J, Chen Y, Wu S, et al., 2015. Comparative transcriptome analysis of latex from rubber tree clone CATAS8-79 and PR107 reveals new cues for the regulation of latex regeneration and duration of latex flow[J]. BMC Plant Biology, 15 (1): 104.

Chen C M, 2012a. Predictive effects of structural variation on citation counts[J]. Journal of the Association for Information Science and Technology, 63 (3): 431–449.

Chen C M, 2017. Science mapping: a systematic review of the literature[J]. Journal of Data and Information Science, 2 (2): 1–40.

Chen C M, Hu Z, Liu S B, et al., 2012b. Emerging trends in regenerative medicine: a scientometric analysis in CiteSpace[J]. Expert Opinion on Biological Therapy, 12 (5): 593–608.

Chen C M, Leydesdorff L, 2014 .Patterns of connections and movements in dual-map overlays: a new method of publication portfolio analysis[J]. Journal of the Association for Information Science and Technology, 65 (2): 334–351.

Chenal J M, Gauthier C, Chazeau L, et al., 2007. Parameters governing strain induced crystallization in filled natural rubber[J]. Polymer, 48 (23): 6 893–6 901.

Choi S, Nah C, Lee S G, et al., 2003. Effect of filler–filler interaction on rheological behaviour of natural rubber compounds filled with both carbon black and silica[J]. Polymer International, 52 (1): 23–28.

Chrispeels M J, Raikhel N V., 1991. Lectins, lectin genes, and their role in plant defense[J]. The plant cell, 3 (1): 1–9.

Cornish K, 1993. The separate roles of plant cis and trans prenyl transferases in cis-1, 4-polyisoprene biosynthesis[J]. The FEBS Journal, 218 (1): 267–271.

Cornish K, 2001. Similarities and differences in rubber biochemistry among plant species[J]. Phytochemistry, 57 (7): 1 123–1 134.

Czuppon A B, Chen Z, Rennert S, et al., 1993. The rubber elongation factor of rubber trees (*Hevea brasiliensis*) is the major allergen in latex[J]. Journal of Allergy and Clinical Immunology, 92 (5): 690–697.

Das A, Stöckelhuber K W, Jurk R, et al., 2008. Modified and unmodified multiwalled carbon nanotubes in high performance solution-styrene-butadiene and butadiene rubber blends[J]. Polymer，49（24）：5 276–5 283.

Dennis M S，Light D R，1989. Rubber elongation factor from *Hevea brasiliensis*. Identification，characterization，and role in rubber biosynthesis[J]. Journal of Biological Chemistry，264（31）：18 608–18 617.

Diani J，Fayolle B，Gilormini P，2009. A review on the Mullins effect[J]. European Polymer Journal，45（3）：601–612.

Fakhru'l-Razi A，Atieh M A，Girun N，et al.，2006. Effect of multi-wall carbon nanotubes on the mechanical properties of natural rubber[J]. Composite Structures，75（1）：496–500.

Fornes T D，Paul D R，2003. Modeling properties of nylon 6/clay nanocomposites using composite theories[J]. Polymer，44（17）：4 993–5 013.

Frogley M D，Ravich D，Wagner H D，2003. Mechanical properties of carbon nanoparticle-reinforced elastomers[J]. Composites Science and technology，63（11）：1 647–1 654.

Ganter M，Gronski W，Reichert P，et al.，2001. Rubber nanocomposites：morphology and mechanical properties of BR and SBR vulcanizates reinforced by organophilic layered silicates[J]. Rubber Chemistry and Technology，74（2）：221–235.

Geethamma V G，Kalaprasad G，Groeninckx G，et al.，2005. Dynamic mechanical behavior of short coir fiber reinforced natural rubber composites[J]. Composites Part A：Applied Science and Manufacturing，36（11）：1 499–1 506.

Gopalan Nair K，Dufresne A，2003. Crab shell chitin whisker reinforced natural rubber nanocomposites. 2. Mechanical behavior[J]. Biomacromolecules，4（3）：666–674.

Hashim A S，Azahari B，Ikeda Y，et al.，1998. The effect of bis（3-triethoxysilylpropyl）tetrasulfide on silica reinforcement of styrene-butadiene rubber[J]. Rubber Chemistry and Technology，71（2）：289–299.

Herman I，Melançon G，Marshall M S，2000. Graph visualization and navigation in information visualization：a survey[J]. IEEE Transactions on Visualization and Computer Graphics，6（1）：24–43.

Hillebrand A，Post J J，Wurbs D，et al.，2012. Down-regulation of small rubber particle protein expression affects integrity of rubber particles and rubber content in *Taraxacum brevicorniculatum*[J]. PLoS One，7（7）：e41874.

Huneau B，2011. Strain-induced crystallization of natural rubber：a review of X-ray diffraction investigations[J]. Rubber Chemistry and Technology，84（3）：425–452.

Ikeda Y，Kameda Y，2004. Preparation of "green" composites by the sol-gel process：in situ silica filled natural rubber[J]. Journal of Sol-Gel Science and Technology，31（1-3）：137–142.

Imbernon L, Norvez S, 2016. From landfilling to vitrimer chemistry in rubber life cycle[J]. European Polymer Journal, 82: 347-376.

Ismail H, Shuhelmy S, Edyham M R, 2002. The effects of a silane coupling agent on curing characteristics and mechanical properties of bamboo fibre filled natural rubber composites[J]. European Polymer Journal, 38（1）: 39-47.

Jacob M, Thomas S, Varughese K T, 2004. Mechanical properties of sisal/oil palm hybrid fiber reinforced natural rubber composites[J]. Composites Science and Technology, 64（7）: 955-965.

Johnson B, Shneiderman B, 1991. Tree-maps: a space-filling approach to the visualization of hierarchical information structures[J]. Proceedings of the 2nd Conference on Visualization'91. IEEE Computer Society Press, 284-291.

Joly S, Garnaud G, Ollitrault R, et al., 2002. Organically modified layered silicates as reinforcing fillers for natural rubber[J]. Chemistry of Materials, 14（10）: 4 202-4 208.

Kang H, Kang M Y, Han K H, 2000a. Identification of natural rubber and characterization of rubber biosynthetic activity in fig tree[J]. Plant Physiology, 123（3）: 1 133-1 142.

Kang H, Kim Y S, Chung G C, 2000b. Characterization of natural rubber biosynthesisin Ficus benghalensis[J]. Plant Physiology and Biochemistry, 38（12）: 979-987.

Kawahara S, Kawazura T, Sawada T, et al., 2003. Preparation and characterization of natural rubber dispersed in nano-matrix[J]. Polymer, 44（16）: 4 527-4 531.

Kelly K J, Kurup V, Zacharisen M, et al., 1993. Skin and serologic testing in the diagnosis of latex allergy[J]. Journal of Allergy and Clinical Immunology, 91（6）: 1 140-1 145.

Kim H, Abdala A A, Macosko C W, 2010. Graphene/polymer nanocomposites[J]. Macromolecules, 43（16）: 6 515-6 530.

Ko J H, Chow K S, Han K H, 2003. Transcriptome analysis reveals novel features of the molecular events occurring in the laticifers of *Hevea brasiliensis*（para rubber tree）[J]. Plant Molecular Biology, 53（4）: 479-492.

Kohjiya S, Ikeda Y, 2000. Reinforcement of general-purpose grade rubbers by silica generated in situ[J]. Rubber Chemistry and Technology, 73（3）: 534-550.

Kohjiya S, Murakami K, Iio S, et al., 2001. In situ filling of silica onto "green" natural rubber by the sol-gel process[J]. Rubber Chemistry and Technology, 74（1）: 16-27.

Kueseng K, Jacob K I, 2006. Natural rubber nanocomposites with SiC nanoparticles and carbon nanotubes[J]. European Polymer Journal, 42（1）: 220-227.

Lagier F, Vervloet D, Lhermet I, et al., 1992. Prevalence of latex allergy in operating room nurses[J]. Journal of Allergy and Clinical Immunology, 90（3）: 319-322.

Leblanc J L，2002. Rubber–filler interactions and rheological properties in filled compounds[J]. Progress in Polymer Science，27（4）：627−687.

Light D R，Dennis M S，1989. Purification of a prenyltransferase that elongates cis-polyisoprene rubber from the latex of *Hevea brasiliensis*[J]. Journal of Biological Chemistry，264（31）：18 589−18 597.

Liu C，Shao Y，Jia D，2008. Chemically modified starch reinforced natural rubber composites[J]. Polymer，49（8）：2 176−2 181.

Lopez D，Amira M B，Brown D，et al.，2016. The *Hevea brasiliensis* XIP aquaporin subfamily：genomic，structural and functional characterizations with relevance to intensive latex harvesting[J]. Plant Molecular Biology，91（4-5）：375−396.

López-Manchado M A，Biagiotti J，Valentini L，et al.，2004. Dynamic mechanical and Raman spectroscopy studies on interaction between single-walled carbon nanotubes and natural rubber[J]. Journal of Applied Polymer Science，92（5）：3 394−3 400.

Meera A P，Said S，Grohens Y，et al.，2009. Nonlinear viscoelastic behavior of silica-filled natural rubber nanocomposites[J]. The Journal of Physical Chemistry C，113（42）：17 997−18 002.

Michael T，Niggemann B，Moers A，et al.，1996. Risk factors for latex allergy in patients with spina bifida[J]. Clinical & Experimental Allergy，26（8）：934−939.

Moneret-Vautrin D A，Beaudouin E，Widmer S，et al.，1993. Prospective study of risk factors in natural rubber latex hypersensitivity[J]. Journal of Allergy and Clinical Immunology，92（5）：668−677.

Nakason C，Kaesaman A，Supasanthitikul P，2004. The grafting of maleic anhydride onto natural rubber[J]. Polymer Testing，23（1）：35−41.

Nakason C，Wannavilai P，Kaesaman A，2006. Effect of vulcanization system on properties of thermoplastic vulcanizates based on epoxidized natural rubber/polypropylene blends[J]. Polymer Testing，25（1）：34−41.

Neuhaus J M，Sticher L，Meins F J R，et al.，1991. A short C-terminal sequence is necessary and sufficient for the targeting of chitinases to the plant vacuole[J]. Proceedings of the National Academy of Sciences，88（22）：10 362−10 366.

Oh S K，Kang H，Shin D H，et al.，1999. Isolation，characterization，and functional analysis of a novel cDNA clone encoding a small rubber particle protein from *Hevea brasiliensis*[J]. Journal of Biological Chemistry，274（24）：17 132−17 138.

Park S J，Cho K S，2003. Filler–elastomer interactions：influence of silane coupling agent on crosslink density and thermal stability of silica/rubber composites[J]. Journal of Colloid and

Interface Science, 267（1）: 86–91.

Poompradub S, Tosaka M, Kohjiya S, et al., 2005. Mechanism of strain-induced crystallization in filled and unfilled natural rubber vulcanizates[J]. Journal of Applied Physics, 97（10）: 103 529.

Potts J R, Shankar O, Murali S, et al., 2013. Latex and two-roll mill processing of thermally-exfoliated graphite oxide/natural rubber nanocomposites[J]. Composites Science and Technology, 74: 166–172.

Praveen S, Chattopadhyay P K, Albert P, et al., 2009. Synergistic effect of carbon black and nanoclay fillers in styrene butadiene rubber matrix: development of dual structure[J]. Composites Part A: Applied Science and Manufacturing, 40（3）: 309–316.

Rajasekar R, Pal K, Heinrich G, et al., 2009. Development of nitrile butadiene rubber–nanoclay composites with epoxidized natural rubber as compatibilizer[J]. Materials & Design, 30（9）: 3 839–3 845.

Rattanasom N, Saowapark T, Deeprasertkul C, 2007. Reinforcement of natural rubber with silica/carbon black hybrid filler[J]. Polymer Testing, 26（3）: 369–377.

Ray S S, Okamoto M, 2003. Polymer/layered silicate nanocomposites: a review from preparation to processing[J]. Progress in Polymer Science, 28（11）: 1 539–1 641.

Sastre J, Fernández-Nieto M, Rico P, et al., 2003. Specific immunotherapy with a standardized latex extract in allergic workers: a double-blind, placebo-controlled study[J]. Journal of Allergy and Clinical Immunology, 111（5）: 985–994.

Thiraphattaraphun L, Kiatkamjornwong S, Prasassarakich P, et al., 2001. Natural rubber-g-methyl methacrylate/poly（methyl methacrylate）blends[J]. Journal of Applied Polymer Science, 81（2）: 428–439.

Shanmugharaj A M, Bae J H, Lee K Y, et al., 2007. Physical and chemical characteristics of multiwalled carbon nanotubes functionalized with aminosilane and its influence on the properties of natural rubber composites[J]. Composites Science and Technology, 67（9）: 1 813–1 822.

Sirisinha C, Saeoui P, Guaysomboon J, 2002. Mechanical properties, oil resistance, and thermal aging properties in chlorinated polyethylene/natural rubber blends[J]. Journal of Applied Polymer Science, 84（1）: 22–28.

Sirisinha C, Saeoui P, Guaysomboon J, 2004. Oil and thermal aging resistance in compatibilized and thermally stabilized chlorinated polyethylene/natural rubber blends[J]. Polymer, 45（14）: 4 909–4 916.

Sussman G L, Tarlo S, Dolovich J, 1991. The spectrum of IgE-mediated responses to latex[J]. Jama，265（21）：2 844–2 847.

Tanaka Y，2001. Structural characterization of natural polyisoprenes：solve the mystery of natural rubber based on structural study[J]. Rubber Chemistry and Technology，74（3）：355–375.

Tang Z，Zhang L，Feng W，et al.，2014. Rational design of graphene surface chemistry for high-performance rubber/graphene composites[J]. Macromolecules，47（24）：8 663–8 673.

Tarachiwin L，Sakdapipanich J，Ute K，et al.，2005. Structural characterization of α -terminal group of natural rubber. 2. Decomposition of branch-points by phospholipase and chemical treatments[J]. Biomacromolecules，6（4）：1 858–1 863.

Toki S，Che J，Rong L，et al.，2013. Entanglements and networks to strain-induced crystallization and stress–strain relations in natural rubber and synthetic polyisoprene at various temperatures[J]. Macromolecules，46（13）：5 238–5 248.

Toki S，Fujimaki T，Okuyama M，2000. Strain-induced crystallization of natural rubber as detected real-time by wide-angle X-ray diffraction technique[J]. Polymer，41（14）：5 423–5 429.

Toki S，Sics I，Ran S，et al.，2002. New insights into structural development in natural rubber during uniaxial deformation by in situ synchrotron X-ray diffraction[J]. Macromolecules，35（17）：6 578–6 584.

Tosaka M，Murakami S，Poompradub S，et al.，2004. Orientation and crystallization of natural rubber network as revealed by WAXD using synchrotron radiation[J]. Macromolecules，37（9）：3 299–3 309.

Trabelsi S，Albouy P A，Rault J，2003. Crystallization and melting processes in vulcanized stretched natural rubber[J]. Macromolecules，36（20）：7 624–7 639.

Turjanmaa K，1987. Incidence of immediate allergy to latex gloves in hospital personnel[J]. Contact Dermatitis，17（5）：270–275.

Valuer P，Balland S，Harf R，et al.，1995. Identification of profilin as an IgE-binding component in latex from *Hevea brasiliensis*：clinical implications[J]. Clinical & Experimental Allergy，25（4）：332–339.

van Beilen J B，Poirier Y，2007. Establishment of new crops for the production of natural rubber[J]. Trends in Biotechnology，25（11）：522–529.

Vandenplas O，Jamart J，Delwiche J P，et al.，2002. Occupational asthma caused by natural rubber latex：outcome according to cessation or reduction of exposure[J]. Journal of Allergy and Clinical Immunology，109（1）：125–130.

Varghese S，Karger-Kocsis J，2003. Natural rubber-based nanocomposites by latex compounding with layered silicates[J]. Polymer，44（17）：4 921–4 927.

Wang Y，Zhang H，Wu Y，et al.，2005. Preparation and properties of natural rubber/rectorite nanocomposites[J]. European Polymer Journal，41（11）：2 776–2 783.

Wagner B，Krebitz M，Buck D，et al.，1999. Cloning，expression，and characterization of recombinant Hev b 3，a *Hevea brasiliensis* protein associated with latex allergy in patients with spina bifida[J]. Journal of Allergy and Clinical Immunology，104（5）：1 084–1 092.

Wei F，Luo S，Zheng Q，et al.，2015. Transcriptome sequencing and comparative analysis reveal long-term flowing mechanisms in *Hevea brasiliensis* latex[J]. Gene，556（2）：153–162.

Wu J，Xing W，Huang G，et al.，2013. Vulcanization kinetics of graphene/natural rubber nanocomposites[J]. Polymer，54（13）：3 314–3 323.

Yassin M S，Lierl M B，Fischer T J，et al.，1994. Latex allergy in hospital employees[J]. Annals of Allergy，72（3）：245–249.

Yunginger J W，Jones R T，Fransway A F，et al.，1994. Extractable latex allergens and proteins in disposable medical gloves and other rubber products[J]. Journal of Allergy and Clinical Immunology，93（5）：836–842.

Zhan Y，Lavorgna M，Buonocore G，et al.，2012. Enhancing electrical conductivity of rubber composites by constructing interconnected network of self-assembled graphene with latex mixing[J]. Journal of Materials Chemistry，22（21）：10 464–10 468.

Zhan Y，Wu J，Xia H，et al.，2011. Dispersion and exfoliation of graphene in rubber by an ultrasonically-assisted latex mixing and in situ reduction process[J]. Macromolecular Materials and Engineering，296（7）：590–602.

Zhang X，Loo L S，2009. Study of glass transition and reinforcement mechanism in polymer/layered silicate nanocomposites[J]. Macromolecules，42（14）：5 196–5 207.

Zhang X，Wang J，Jia H，et al.，2016. Multifunctional nanocomposites between natural rubber and polyvinyl pyrrolidone modified graphene[J]. Composites Part B：Engineering，84：121–129.

8 天然橡胶学科领域前沿热点分析

8.1 引言

巴西橡胶树（又称橡胶树）由于其产量高、品质好、经济寿命长、生产成本低等优点，成为人工栽培中最为重要的产胶植物，其产量占世界天然橡胶总产量的99%以上。橡胶是热带地区典型的经济林作物，是重要的战略物资。在热带农林业中，橡胶具有特殊和重要的地位。迄今为止，天然橡胶在航天、军工及医疗等高端和特殊用途领域中仍具有不可替代性。当前，天然橡胶产业持续低迷，国际天然橡胶产品供大于求，但我国天然橡胶的自给率仍不到20%，国内高端和特殊用途的高性能胶几乎完全依赖进口（国家天然橡胶产业技术体系，2016；International Rubber Research and Development Board，2006）。我国天然橡胶产业发展中的一些重大问题已逐渐发生转变，如从早期的追求高产转变为高产与优质并重、胶木兼优品种的选育和推广、劳动力成本的不断上升、加工领域工艺改进和技术创新等。从我国天然橡胶产业发展历程来看，科学技术的提升是推动橡胶产业升级的重要动力。当前一些高新尖的技术领域，如天然高分子或纳米微粒补强天然橡胶合成纳米复合材料、橡胶微生物降解、产胶植物橡胶生物合成与调控等，未来很可能极大地影响天然橡胶产业的发展。

研究前沿是科技创新过程中最新、最有潜力、最具前瞻性和引领性的研究方向。有效识别领域发展前沿，可以对未来的研究趋势作出有效预判，从而将人力、物力和财力精准投入到最具战略研究价值的科技前沿（Price，1964；Small and Griffith，1974；王立学和冷伏海，2010；罗瑞等，2018）。学科领域前沿的主要特征有两个：时间上，产生于近期并延伸到未来；创新程度上，蕴含较高创新价值的研究。值得注意的是，科技创新中的创新程度有着巨大区别，依据创新程度的高低又可以将研究前沿分为渐进性创新和变革性创新（Persson，1994；Morris et al.，2003）。科学知识图谱是进行领域分析和可视化的通用过程，其分析范围可以是一门学科、一个研究领域或特定研究问题的主题领域。换句话说，知识图谱的分析单元是科学知识的一个领域，它通过一个科学团体或更精确定义的专业成员的智力贡献集合来反映（Chen，2006；2017）。常用的科学文献来源有Web of Science、Scopus、Google Scholar和PubMed等。科学计量学方法包括作者共被引分析、文献共被引分析、共词分析和结构变异分析等（Chen，2006，2017；White and McCain，1998；Callon et al.，1983）。知识图谱工具通常将一组文献作为输入，生成具有复杂结构的交互式图像用于定量分析和视觉探索。许多知识图谱技术起源于共被引分析理论，这一理论描述了知识基础在共被引文献网络中的结构特征（Chen，2006，2017；White and McCain，1998；Callon et al.，1983；Shneider，2009；Fuchs，1993）。本章依据科学计量学理论，在前一章识别研究前沿（渐进性创新）和演进历程的基础上，基于CiteSpace可视化分析工具，对2004—2017年天然橡胶学科领域的研究成果和重要文献进行识别和可视化，建立学科领域知识图谱，跟踪和揭示国际天然橡胶学科领域的前沿热点（变革性创新），以期为科研管理者、政策制定者、相关科研人员及天然橡胶产业发展提供科学参考和借鉴。

8.2 数据来源与研究方法

8.2.1 数据来源

数据来源于Web of Science核心合集的SCI-EXPANDED和SSCI。本章定义的天然橡胶，是指从巴西橡胶树（*Hevea brasiliensis*）提取的天然胶乳。为了确保天然橡胶文献数据集的查全率，采用表8-1制定的检索策略进行主题检索。检索词主要有天然橡胶、天然胶乳、橡胶胶乳、橡胶树、橡胶林、橡胶种植园、胶园、植胶区、橡胶间作、橡胶生物合成、橡胶产量、割胶、产胶、排胶等。经检索得到9 960条记录，除重后获得9 579篇文献，这些文献共引用了179 317篇引文（检索日期：2019-09-17，数据库更新日期：2019-09-16）。

表8–1　天然橡胶文献数据集的主题检索策略

Tab. 8–1　Topic search queries used for data collection in natural rubber literature

数据集 Set	检索策略 Topic search
1	主题=（“rubber tree*” or “*Hevea*” or “natur* rubber” or “natur* latex” or “nr latex” or “rubber latex” or “lactiferous plant” or “rubber yielding plant” or “rubber producing plant*” or “rubber producing crop*” or “rubber biosynthesis”）and 文献类型=（Article or Review）and 语种=（English）；索引=SCI-EXPANDED，SSCI；年份=2004–2018。
2	主题=（“rubber”）and主题=（“tapping” or “plantation*” or “yard” or “yield” or “productivity” or “garden” or “forest” or “NR” or “planting area*” or “planting” or “growing area*” or “growing state” or “intercrop*” or “interplanting” or “latex production” or “latex drainage”）and 文献类型=（Article or Review）and 语种=（English）；索引=SCI-EXPANDED，SSCI；年份=2004–2018。
3	主题=1 or 2；索引=SCI-EXPANDED，SSCI；年份=2004–2018。

8.2.2　研究方法

采用CiteSpace（5.5.R2）进行天然橡胶文献数据集的可视化分析。CiteSpace使用时间切片技术构建随时间变化的时间序列网络模型，并综合这些单独的网络形成一个概览网络，以便系统地回顾相关文献。以每年引用次数前100的文献构建当年共被引网络，然后合成各个网络。合成的网络被划分为多个共被引文献聚类，相似的论文和相关的聚类被定位在接近的位置，而不同论文和聚类则相距较远。采用Chen的方法，排除检索结果中的干扰文献（Chen，2017），基于CiteSpace技术特点，对检索策略的持续精炼以及对检索结果文献的手工剔除，在一定程度上会导致相关研究文献的缺失并影响文献的关联。通过对原始检索结果所生成的图谱，辨别分析有效文献所生成的大型活跃聚类，而对无效文献生成的聚类则加以识别和排除，能够有效地保证文献查全率，同时能够排除与研究无关的干扰文献。每个聚类成员（被引文献）代表研究领域的知识基础，引用这些文献的施引文献是与这些聚类相关的研究前沿（Chen，2006；2017）。本章只列出前5～10篇被引文献和施引文献进行陈述和解读。一个节点的引文历史描述为若干个引用年轮，每一个引用年轮代表共被引网络中相应年份的引用次数。

8.3 天然橡胶文献共被引网络学科领域总体概况

合成的网络包含1 499篇引文。4个最大的连接网络包括1 272个节点，占整个网络的84%。该网络具有非常高的模块化值，为0.830 7，表明各学科领域在共被引聚类中有明确的定义。图谱展示了4个主要学科领域（图8-1），左侧区域涉及化学与材料科学，右上方涉及生物科学、生态与环境科学，右下方涉及免疫学。不同颜色区域表示这些区域共被引连接首次出现的时间，紫色区域比粉色区域生成的时间早，黄色区域是在粉色区域之后生成的。黄色区域是仍在持续发展的聚类，目前仍在活跃的大型聚类象征着学科的前沿方向，也是本章重点分析的部分。每个聚类都可以通过标题术语、关键字和引用聚类文献的抽象术语进行标记，如最右侧黄色区域被标记为橡胶人工林与生态环境，表明关于橡胶人工林与生态环境的论文引用了#8聚类。表8-2按照核心论文数量列出了前13个聚类。轮廓值是衡量聚类同质性或一致性的指标，同质聚类的平均轮廓值趋于1（Chen，2006；2017）。

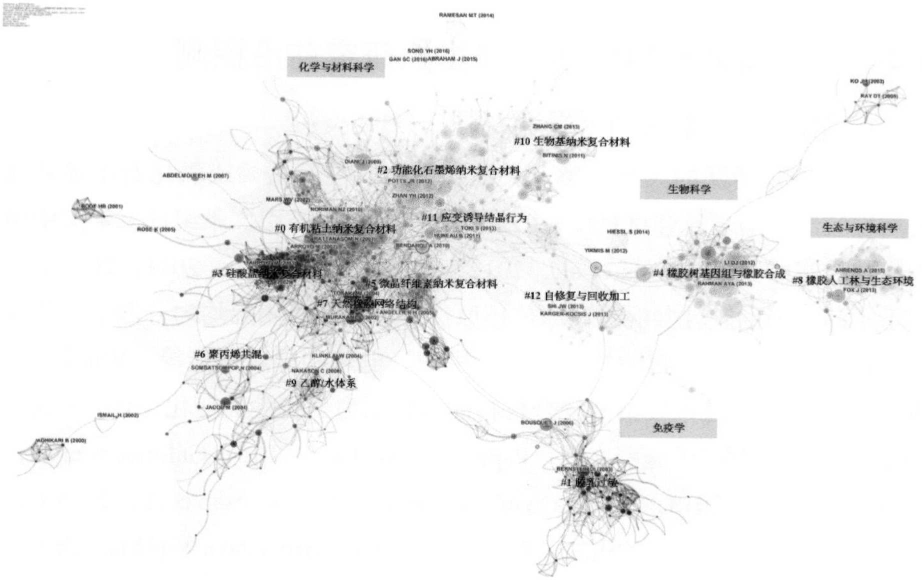

图8-1 天然橡胶文献共被引网络

Fig. 8-1 Natural rubber research shown in a network of co-cited references

表8-2 天然橡胶文献共被引聚类所属学科领域及研究前沿

Tab. 8-2 Discipline field and research fronts in natural rubber clusters of co-cited references

学科领域 Discipline field	研究前沿 Research fronts	核心论文 Core papers	轮廓值 Silhouette	平均年 Mean year	聚类 Cluster
化学与材料科学	有机黏土纳米复合材料	164	0.680	2008	0
免疫学	乳胶过敏	128	0.997	2003	1
化学与材料科学	功能化石墨烯纳米复合材料	121	0.846	2013	2
化学与材料科学	硅酸盐纳米复合材料	120	0.884	2003	3
生物科学	橡胶树基因组与橡胶合成	105	0.951	2010	4
化学与材料科学	微晶纤维素纳米复合材料	90	0.898	2007	5
化学与材料科学	聚丙烯共混	85	0.902	2004	6
化学与材料科学	天然橡胶网络结构	74	0.867	2004	7
生态与环境科学	橡胶人工林与生态环境	61	0.989	2012	8
化学与材料科学	乙醇/水体系	61	0.937	2005	9
化学与材料科学	生物基纳米复合材料	50	0.967	2012	10
化学与材料科学	应变诱导结晶行为	50	0.912	2011	11
化学与材料科学	自修复与回收加工	41	0.964	2013	12

8.4 化学与材料科学活跃聚类的研究前沿探测

8.4.1 功能化石墨烯纳米复合材料——聚类#2

聚类#2是活跃的大型聚类，关注功能化改性石墨烯或氧化石墨烯补强天然橡胶制备纳米复合材料（表8-3），由121篇共被引文献组成。聚类中引用次数较高的文章（Potts et al.，2012，2013；Zhan et al.，2011，2012），展示了超声辅助胶乳共混法和原位还原法等将氧化石墨烯均匀分散于天然橡胶基体，显著提高材料的拉伸强度、机械、电热和屏蔽性能等。Wu等系统研究了石墨烯/天然橡胶纳米复合材料的硫化动力学特性的变化（Wu et al.，2013）。覆盖率前5的施引文献（Papageorgiou et al.，2015；Galimberti et al.，2014；Wu et al.，2013；Srivastava and Mishra，2018；Mensah et al.，2018），引用了该聚类8%～12%的引文。Papageorgiou等、Srivastava等和Mensah等的文章综述了不同类型石墨烯应用于补强天然橡胶或弹性体纳米复合材料及其对复合材料的拉伸强度、热稳定性、气体屏蔽、电学、机械和动态力学性能等的影响（Papageorgiou et al.，2015；Srivastava and Mishra，2018；Mensah et al.，2018）。其他施引文献也在关注改进的方法制备功能化改性石墨烯或氧化石墨烯/天然橡胶纳米复合材料及其性能的增强（Galimberti et al.，2014；Wu et al.，2013）。

表8-3 功能化石墨烯纳米复合材料的被引文献和施引文献——聚类#2

Tab. 8-3 Cited references and citing articles of functionalized graphene nanocomposites-cluster #2

被引文献Cited references		施引文献Citing articles	
引用次数 Cites	作者（年）期刊，卷，页 Author（Year）Journal，Volume，Page	覆盖率（%）Coverage（%）	作者（年）Author（Year）
74	Potts J R（2012）Macromolecules，V45，P6045	12	Papageorgiou，Dimitrios G（2015）
51	Zhan Y H（2012）J Mater Chem，V22，P10464	9	Galimberti，Maurizio（2014）
51	Potts J R（2013）Compos Sci Technol，V74，P166	8	Wu S W（2013）
50	Zhan Y H（2011）Macromol Mater Eng，V296，P590	8	Srivastava，Suneel Kumar（2018）
48	Wu J R（2013）Polymer，V54，P3314	8	Mensah，Bismark（2018）

注：被引文献的引用次数是本章构建的天然橡胶文献共被引网络中累计的引用次数。下同

Note: The citations of cited references are the cumulative citations in the co-citation network of natural rubber references constructed in this chapter. The same below

8.4.2 生物基纳米复合材料——聚类#10

聚类#10代表了聚乳酸、纳米微晶纤维素等生物基体补强天然橡胶制备纳米复合材料活跃的施引文献和被引文献（表8-4）。该聚类引用次数第一的文章由Bitinis等发表，关于聚乳酸/天然橡胶共混物的微观结构、结晶行为和机械性能研究（Bitinis et al.，2011）。其他共被引文献代表了聚乳酸基或形状记忆聚乳酸基与天然橡胶或环氧化天然橡胶制备共混物及其韧性、晶体稳定性增强的知识基础（Zhang et al.，2013；Heuwers et al.，2013；Jaratrotkamjorn et al.，2012；Xu et al.，2016）。覆盖率前5的施引文献，引用了该聚类6%～7%的引文（Cao et al.，2018a，2018b；Heuwers et al.，2013a；Quitmann et al.，2013；Chen et al.，2014）。Cao等采用两种方法制备海鞘纳米微晶纤维素/天然橡胶纳米复合材料，对其形态、力学性能和水膨胀行为进行了比较研究（Cao et al.，2018a；2018b）。Heuwers等、Quitmann等研究了不同形状记忆聚乳

酸基天然橡胶纳米复合材料的储能、力学和机械应力性能（Heuwers et al.，2013a；Quitmann et al.，2013）。Chen等研发了一种生物基动态硫化聚乳酸/天然橡胶共混物，其中交联NR相具有连续的网状分散（Chen et al.，2014）。

表8-4　生物基纳米复合材料的被引文献和施引文献——聚类#10

Tab. 8-4　Cited references and citing articles of biobased nanocomposites–cluster #10

被引文献Cited references		施引文献Citing articles	
引用次数 Cites	作者（年）期刊，卷，页 Author（Year）Journal，Volume，Page	覆盖率（%） Coverage（%）	作者（年） Author（Year）
35	Bitinis N（2011）Mater Chem Phys，V129，P823	7	Cao L M（2018a）
27	Zhang C M（2013）Mater Design，V45，P198	6	Heuwers，Benjamin（2013a）
25	Heuwers B（2013b）Macromol Rapid Comm，V34，P180	6	Quitmann，Dominik（2013）
25	Jaratrotkamjorn R（2012）J Appl Polym Sci，V124，P5027	6	Chen Y K（2014）
23	Xu C H（2016）ACS APPL Mater Inter，V8，P17728	6	Cao L M（2018b）

8.4.3　应变诱导结晶行为——聚类#11

聚类#11的核心成员揭示了天然橡胶或弹性体纳米复合材料的应变诱导结晶行为（表8-5）。共被引文献中（Toki et al.，2013；Huneau，2011；Candau et al.，2014；Brüning et al.，2012；Vieyres et al.，2013），Toki等通过原位同步广角X射线衍射法研究硫化和未硫化天然橡胶、异戊橡胶在单轴变形过程中的应力—应变关系（Toki et al.，2013）。Huneau将X射线衍射法用于天然橡胶结晶与机械响应之间的关系研究及对交联密度、炭黑填料、温度、应变速率等不同因素的影响（Huneau，2011）。施引文献中，覆盖率前5的文献（Fu et al.，2015；Rublon et al.，2013；Pérez-Aparicio et al.，2013a，2013b；Mondal et al.，2013），引用了该聚类4%～5%的引文。Fu等利用同步加速广角X射线衍射研究有机改性蒙脱土填充聚异戊橡胶的应变诱导结晶行为（Fu et al.，2015）。Rublon等开发出一种将同步辐射与自制疲劳拉伸机器耦合的试验装置，该装置用于测量在不同载荷条件下炭黑填充天然橡胶的疲劳裂纹周围的结晶度分布（Rublon et al.，2013）。其他施引文献侧重基于机械响应、核磁共振、X射线散射等不同技术组合测量天然橡胶体系的拉伸、形变过程中的应变结晶响应（Pérez-Aparicio et al.，2013a，2013b；Mondal et al.，2013）。

表8-5 应变诱导结晶行为的被引文献和施引文献——聚类#11

Tab. 8-5 Cited references and citing articles of strain-induced crystallization behavior-cluster #11

被引文献Cited references		施引文献Citing articles	
引用次数 Cites	作者（年）期刊，卷，页 Author（Year）Journal，Volume，Page	覆盖率(%) Coverage(%)	作者（年） Author（Year）
34	Toki S（2013）Macromolecules，V46，P5238	5	Fu X A（2015）
32	Huneau B（2011）Rubber Chem Technol，V84，P425	5	Rublon，Pierre（2013）
31	Candau N（2014）Macromolecules，V47，P5815	5	Pérez-Aparicio，Roberto（2013a）
31	Brüning K（2012）Macromolecules，V45，P7914	4	Pérez-Aparicio，Roberto（2013b）
28	Vieyres A（2013）Macromolecules，V46，P889	4	Mondal，Manas（2013）

8.4.4 自修复与回收加工——聚类#12

聚类#12是最近形成的聚类，代表废旧橡胶的自修复与回收加工，包含41篇共被引文献，前5篇施引文献引用了该聚类6%～15%的引文（表8-6）。引用频次较高的论文涉及不同方法回收橡胶、影响橡胶回收的因素及再生橡胶的结构、性能、加工和应用（Karger-Kocsis et al.，2013；Shi et al.，2013；Faruk et al.，2012；Riyajan et al.，2012；Myhre et al.，2012）。Imbernon等概述了在热、氧、机械和化学试剂等的作用下，使硫化橡胶中的三维网络发生降解，得到具有流动性的再生橡胶。还提出一种基于环氧化天然橡胶的模型体系，能够区分和量化应力、松弛、溶胀和黏附各过程，对固有松弛具有玻璃化的化学优势（Imbernon et al.，2016a，2016b）。其他施引文献（Xiang et al.，2016；Denissen et al.，2016；Xu et al.，2016），如Xiang等引入有机催化剂甲基丙烯酸铜（MA-Cu）与氯丁橡胶共混，发现MA-Cu能够激发橡胶网络中二硫键的交换反应（Xiang et al.，2016）。Xu等将羧基化丁苯橡胶与氧化锌共混，提高离子键浓度，提高橡胶拉伸强度（Xu et al.，2016）。

表8-6　自修复与回收加工的被引文献和施引文献——聚类#12

Tab. 8-6　Cited references and citing articles of self-healing and recycling-cluster #12

被引文献Cited references		施引文献Citing articles	
引用次数 Cites	作者（年）期刊，卷，页 Author（Year）Journal，Volume，Page	覆盖率（%） Coverage（%）	作者（年） Author（Year）
29	Karger-kocsis J（2013）J Mater Sci，V48，P1	15	Imbernon，Lucie（2016a）
18	Shi J W（2013）J Appl Polym Sci，V129，P999	7	Imbernon，Lucie（2016b）
17	Faruk O（2012）Prog Polym Sci，V37，P1552	7	Xiang H P（2016）
12	Riyajan S A（2012）Carbohyd Polym，V89，P251	6	Denissen，Wim（2016）
11	Myhre M（2012）Rubber Chem Technol，V85，P408	6	Xu C H（2016）

8.5　生物科学活跃聚类的研究前沿探测

橡胶树基因组与橡胶合成——聚类#4

聚类#4是较活跃的大型聚类（表8-7），内容主要涉及橡胶树全基因组测序及调控橡胶生物合成途径与关键基因等研究，包括105个聚类成员。该聚类中引用次数最高的是Rahman等初步构建橡胶树基因组序列框架图，同时鉴定了与橡胶生物合成、橡胶木材形成、抗病性和致敏性相关的关键基因（Rahman et al.，2013）。其他引文（Li et al.，2012；Tang et al.，2016；Triwitayakorn et al.，2011；Chow et al.，2012，2007；Nawamawat et al.，2011；Sansatsadeekul et al.，2011；Berthelot et al.，2014；Hillebrand et al.，2012），如Li等、Triwitayakorn等开发大量EST-SSR标记，广泛应用于橡胶树标记辅助选择、遗传多样性、DNA指纹图谱、遗传图谱等（Li et al.，2012；Triwitayakorn et al.，2011）。Tang与Hu等2016年合作的论文发布了橡胶树基因组草图，发现REF/SRPP基因家族的几个成员（Tang et al.，2016）。Chow等的文章与橡胶生物合成关系密切，发现了MEP途径（Chow et al.，2012）。Berthelot等综述了与橡胶树橡胶粒子相关的两种重要蛋白的进展：橡胶延伸因子（REF）和小橡胶粒子蛋白（SRPP）（Berthelot et al.，2014）。施引文献包括研究性论文和综述（Duan et al.，2013；Tang et al.，2010；Dusotoit-Coucaud et al.，2010a，2010b；Deng et al.，2018；Montoro et al.，2018；de Faÿ et al.，2011；Zou et al.，2015），引用覆盖率较高的是Duan等通过转录组测序在橡胶树叶片、树皮、胶乳、胚性组织和根中鉴定AP2/ERF（乙烯反应因子）的超家族（Duan et al.，2013）。Deng等揭示了茉莉酸在橡胶树乳管细胞中调控橡胶生物合成的信号通路研究（Deng et al.，2018）。

表8-7 橡胶树基因组与橡胶合成的被引文献和施引文献——聚类#4

Tab. 8-7 Cited references and citing articles of genome in rubber tree and rubber biosynthesis-cluster #4

被引文献Cited references		施引文献Citing articles	
引用次数 Cites	作者（年）期刊，卷，页 Author（Year）Journal，Volume，Page	覆盖率(%) Coverage(%)	作者（年） Author（Year）
65	Rahman A Y A（2013）BMC Genomics，V14，P75	7	Duan C F（2013）
50	Li D J（2012）BMC Genomics，V13，P192	7	Rahman，Ahmad Yamin Abdul（2013）
47	Tang C R（2016）Nat Plants，V2，P16073	6	Tang C R（2010）
41	Triwitayakorn K（2011）DNA Res，V18，P471	5	Berthelot，Karine（2014）
40	Chow K S（2012）J Exp Bot，V63，P1863	5	Dusotoit-Coucaud，Anais（2010a）
36	Nawamawat K（2011）Colloid Surface A，V390，P157	5	Dusotoit-Coucaud，Anais（2010b）
35	Chow K S（2007）J Exp Bot，V58，P2429	5	Deng X M（2018）
26	Sansatsadeekul J（2011）J Biosci Bioeng，V111，P628	5	Montoro，Pascal（2018）
25	Berthelot K（2014）Biochimie，V106，P1	4	de Faÿ，Elisabeth（2011）
23	Hillebrand A（2012）PLoS one，V7，e41874	4	Zou Z（2015）

8.6 生态与环境科学活跃聚类的研究前沿探测

橡胶人工林与生态环境——聚类#8

聚类#8也是最近形成的聚类，代表了土地利用变化背景下橡胶林的扩张种植对生态环境的影响，由61篇共被引文献组成（表8-8）。引用次数最高的是Ahrends等2015年发表在Global Environmental Change上的一篇论文，关于东南亚地区橡胶林的迅速扩张种植，对今后环境、生物多样性造成的巨大影响（Ahrends et al.，2015）。其他引文大多数关于利用Landsat影像量化西双版纳土地利用/土地覆盖的变化，最明显的变化是森林覆盖的减少和橡胶林的增加，构建橡胶树和天然林归一化植被指数（NDVI）、增强型植被指数（EVI）、地表水指数（LSWI）和近红外反射率（NIR）的时间剖面图（Fox et al.，2013；Li et al.，2012；Dong et al.，2013；Xu et al.，2014）。Warren-Thomas等综述了对天然橡胶需求的日益增长，导致大面积的森林或湿地转向单一种植的橡胶人工林，对生物多样性带来负面影响（Warren-Thomas et al.，2015）。其他施引文献涉及热带雨林转变为橡胶人工林对土壤肥力、理化性质、碳储量、碳氮含量、土壤微生物活性、生态系统功能和经济效益等的影响（Guillaume et al.，2016a，2016b；Drescher et al.，2016；Clough et al.，2016）。

表8-8　橡胶人工林与生态环境的被引文献和施引文献——聚类#8

Tab. 8-8　Cited references and citing articles of rubber plantations and ecological environment-cluster #8

被引文献Cited references		施引文献Citing articles	
引用次数 Cites	作者（年）期刊，卷，页 Author（Year）Journal，Volume，Page	覆盖率(%) Coverage(%)	作者（年） Author（Year）
52	Ahrends A（2015）Global Environ Chang，V34，P48	5	Warren-Thomas，Eleanor（2015）
51	Fox J（2013）J Peasant Stud，V40，P155	5	Guillaume，Thomas（2016a）
47	Li Z（2012）Appl Geogr，V32，P420	5	Drescher，Jochen（2016）
43	Dong J W（2013）Remote Sens Environ，V134，P392	5	Clough，Yann（2016）
42	Xu J C（2014）Ecol Indic，V36，P749	4	Guillaume，Thomas（2016b）

8.7　基于关键词共现的研究热点分析

前文通过分析活跃聚类的施引文献和引文探测学科的研究前沿，为进一步分析研究热点，本节将利用CiteSpace构建关键词共现网络图谱，根据关键词共现强度确定研究热点。结合前文活跃聚类的时间分布，将时间框定在2010—2018年，时间切片为1年，选取每年出现频次前100的关键词建立时区图谱（图8-2）。2010年以来的研究热点主要集聚在以下几个领域：一是填料补强天然橡胶制备纳米复合材料及其结构和性能的增强。从节点大小来看，关键节点主要集中在2010年，最大节点为“纳米复合材料”。实际上，科研人员对于天然橡胶纳米复合材料的研究一直在持续，主要表现在此后几年相关关键词包括“共混”“化学改性”“表面改性”“石墨烯”“聚乳酸”“聚合物基体”等共现频次均较高，且连线密集。同样复合材料结构和性能强化的研究也一直在持续，“诱导结晶”“机械性能”“微观结构”“动力学性能”等关键词共现增强。二是废旧橡胶自修复和回收加工，主要体现在“轮胎用橡胶”“废轮胎胶粉”“硫化”“疲劳”“老化”“热降解”等关键词的频次增强。在硫化天然橡胶或复合材料的加工和使用过程中，外部刺激如热、应力和照射等对材料不可逆的损伤会造成力学性能下降，使用寿命降低，因此具有自我修复能力的复合材料的合成也成为科研人员研究的热点课题，这与前文的分析一致。三是橡胶树基因组测序、调控橡胶生物合成路径与基因研究。“生物合成”成为该时段突现值最高的关键词，还涉及“基因表达”“乳管”“聚异戊二烯”等。四是橡胶林的扩张对生态环境的影响，包括“多样性”“热带雨林”“土地利用”“西双版纳”等关键词，同时特别关注我国西南地区橡胶人工林替代热带雨林对生态环境造成的影响。

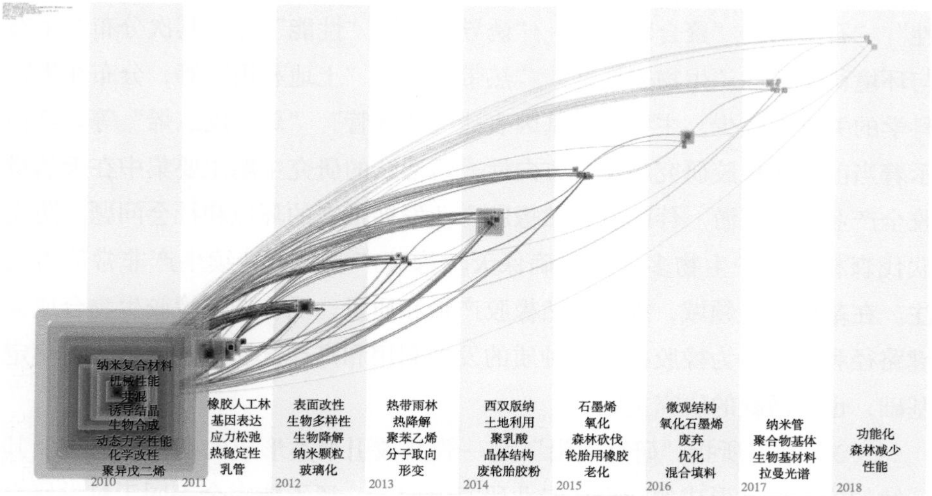

图8–2　天然橡胶关键词共现时区网络及每年高频关键词

Fig. 8–2　Natural rubber research shown in a timezone network of co–word and annual high–frequency keywords

8.8 本章小结

一个领域的研究前沿体现的是该领域当前的科学发展水平。天然橡胶领域的跨学科、跨专业和学科交叉视角明显，使得新兴研究领域和主题不断出现。本章借助知识图谱可视化分析概述了2004—2018年天然橡胶领域重要的基础文献，突出了持续发展的研究领域，并通过关键词共现分析寻找研究热点，探索活跃的研究方向，主要研究结果如下：

（1）在共被引文献和施引文献分析的基础上，按照学科探测研究前沿，在三大学科探测出6个研究前沿。一是化学与材料科学识别出4个研究前沿：功能化改性石墨烯或氧化石墨烯补强天然橡胶制备纳米复合材料；生物基体补强天然橡胶制备纳米复合材料；弹性体或天然橡胶纳米复合材料的应变诱导结晶行为；废旧橡胶的自修复与回收加工。二是生物科学识别出1个研究前沿：橡胶树基因组测序、调控橡胶树乳管橡胶生物合成路径与关键基因。三是生态与环境科学识别出1个研究前沿：土地利用变化背景下橡胶人工林的扩张对生态环境的影响。

（2）在关键词共现频次和强度分析基础上，发现高频和高突现关键词主要分布在化学与材料科学领域，如“纳米复合材料”“共混”“改性”“石墨烯”“聚合物基体”“诱导结晶”“性能”等，其次分布在生态与环境科学，如“生物多样性”“热带雨林”“土地利用”等，分布在生物科学的关键词较少，主要有“基因表达”“乳管”“聚异戊二烯”等。这预示着当前天然橡胶研究趋于多方向性，但活跃的研究主题主要集中在天然橡胶全产业链的下游。伴随着天然橡胶产业发展带来的环境和社会问题，防止砍伐森林，保护生物多样性，确保天然橡胶产业的可持续生产非常值得关注。在基础研究领域，提高天然橡胶产量和质量，明确调控橡胶生物合成关键路径和基因，为橡胶树优异种质的发掘利用和高产优质抗逆遗传改良奠定基础，也是活跃的研究主题。

（3）本书所指“研究前沿”是一簇共被引聚类形成的高被引论文及其后续的施引论文形成的一个“专业研究方向”，还不能完全等同于科学研究中的前沿科学问题和前沿研究领域，所以本方法只是监测分析科学研究发展态势的一种视角。另外，论文的写作、发表和被引用存在一定的滞后性，影响了研究前沿成果的及时揭示，因此需要补充各类相关信息，如施引论文，才能更为全面地监测和分析科学研究发展态势。

参考文献

国家天然橡胶产业技术体系，2016. 中国现代农业产业可持续发展战略研究天然橡胶分册[M]. 北京：中国农业出版社. 11–35.

罗瑞，许海云，董坤，2018. 领域前沿识别方法综述[J]. 图书情报工作，62（23）：119–131.

王立学，冷伏海，2010. 简论研究前沿及其文献计量识别方法[J]. 情报理论与实践，33（3）：54–58.

Ahrends A，Hollingsworth P M，Ziegler A D，et al.，2015. Current trends of rubber plantation expansion may threaten biodiversity and livelihoods[J]. Global Environmental Change，34：48–58.

Berthelot K，Lecomte S，Estevez Y，et al.，2014. *Hevea brasiliensis* REF（Hev b1）and SRPP（Hev b3）：an overview on rubber particle proteins[J]. Biochimie，106：1–9.

Bitinis N，Verdejo R，Cassagnau P，et al.，2011. Structure and properties of polylactide/natural rubber blends[J]. Materials Chemistry and Physics，129（3）：823–831.

Brüning K，Schneider K，Roth S V，et al.，2012. Kinetics of strain-induced crystallization in natural rubber studied by WAXD：dynamic and impact tensile experiments[J]. Macromolecules，45（19）：7 914–7 919.

Callon M，Courtial J P，Turner W A，et al.，1983. From translations to problematic networks：an introduction to co-word analysis[J]. Social Science Information，22（2）：191–235.

Candau N，Laghmach R，Chazeau L，et al.，2014. Strain-induced crystallization of natural rubber and cross-link densities heterogeneities[J]. Macromolecules，47（16）：5 815–5 824.

Cao L M，Huang J R，Chen Y K，2018a. Dual cross-linked epoxidized natural rubber reinforced by tunicate cellulose nanocrystals with improved strength and extensibility[J]. ACS Sustainable Chemistry & Engineering，6（11）：14 802–14 811.

Cao L M，Yuan D S，Fu X F，et al.，2018b. Green method to reinforce natural rubber with tunicate cellulose nanocrystals via one-pot reaction[J]. Cellulose，25（8）：4 551–4 563.

Chen C M，2006. CiteSpace II：Detecting and visualizing emerging trends and transient patterns in scientific literature[J]. Journal of the American Society for Information Science and Technology，57（3）：359–377.

Chen C M，2017. Science mapping：a systematic review of the literature[J]. Journal of Data and Information Science，2（2）：1–40.

Chen Y K，Yuan D S，Xu C H，2014. Dynamically vulcanized biobased polylactide/natural rubber blend material with continuous cross-linked rubber phase[J]. ACS Applied Materials & Interfaces，6（6）：3 811–3 816.

Chow K S，Mat-Isa M N，Bahari A，et al.，2012. Metabolic routes affecting rubber biosynthesis in *Hevea brasiliensis* latex[J]. Journal of Experimental Botany，63（5）：1 863–1 871.

Chow K S，Wan K L，Isa M N M，et al.，2007. Insights into rubber biosynthesis from transcriptome analysis of *Hevea brasiliensis* latex[J]. Journal of Experimental Botany，58（10）：2 429–2 440.

Clough Y，Krishna V V，Corre M D，et al.，2016. Land-use choices follow profitability at the expense of ecological functions in indonesian smallholder landscapes[J]. Nature Communications，7：13 137.

de Faÿ E，2011. Histo-and cytopathology of trunk phloem necrosis，a form of rubber tree（*Hevea brasiliensis* Müll. Arg.）Tapping Panel Dryness[J]. Australian Journal of Botany，59（6）：563–574.

Deng X M，Guo D，Yang S G，et al.，2018. Jasmonate signalling in the regulation of rubber biosynthesis in laticifer cells of rubber tree，*Hevea brasiliensis*[J]. Journal of Experimental Botany，69（15）：3 559–3 571.

Denissen W，Winne J M，Du Prez F E，2016. Vitrimers：permanent organic networks with glass-like fluidity[J]. Chemical Science，7（1）：30–38.

Dong J W，Xiao X M，Chen B Q，et al.，2013. Mapping deciduous rubber plantations through integration of PALSAR and multi-temporal Landsat imagery[J]. Remote Sensing of Environment，134：392–402.

Drescher J，Rembold K，Allen K，et al.，2016. Ecological and socio-economic functions across tropical land use systems after rainforest conversion[J]. Philosophical Transactions of the Royal Society B：Biological Sciences，371（1 694）：20150275.

Duan C F，Argout X，Gébelin V，et al.，2013. Identification of the *Hevea brasiliensis* AP2/ERF superfamily by RNA sequencing[J]. BMC Genomics，14（1）：30.

Dusotoit-Coucaud A，Kongsawadworakul P，Maurousset L，et al.，2010a. Ethylene stimulation of latex yield depends on the expression of a sucrose transporter（*HbSUT1B*）in rubber tree（*Hevea brasiliensis*）[J]. Tree Pysiology，30（12）：1 586–1 598.

Dusotoit-Coucaud A，Porcheron B，Brunel N，et al.，2010b. Cloning and characterization of a new polyol transporter（*HbPLT2*）in *Hevea brasiliensis*[J]. Plant and Cell Physiology，51

（11）：1 878–1 888.

Faruk O，Bledzki A K，Fink H P，et al.，2012. Biocomposites reinforced with natural fibers：2000-2010[J]. Progress in Polymer Science，37（11）：1 552–1 596.

Fox J，Castella J C，2013. Expansion of rubber（*Hevea brasiliensis*）in Mainland Southeast Asia：what are the prospects for smallholders? [J]. The Journal of Peasant Studies，40（1）：155–170.

Fu X，Huang G S，Xie Z T，et al.，2015. New insights into reinforcement mechanism of nanoclay-filled isoprene rubber during uniaxial deformation by in situ synchrotron X-ray diffraction[J]. RSC Advances，5（32）：25 171–25 182.

Fuchs S，1993. A sociological theory of scientific change[J]. Social Forces，71（4）：933–953.

Galimberti M，Cipolletti V，Musto S，et al.，2014. Recent advancements in rubber nanocomposites[J]. Rubber Chemistry and Technology，87（3）：417–442.

Guillaume T，Holtkamp A M，Damris M，et al.，2016a. Soil degradation in oil palm and rubber plantations under land resource scarcity[J]. Agriculture，Ecosystems & Environment，232：110–118.

Guillaume T，Maranguit D，Murtilaksono K，et al.，2016b. Sensitivity and resistance of soil fertility indicators to land-use changes：new concept and examples from conversion of Indonesian rainforest to plantations[J]. Ecological Indicators，67：49–57.

Heuwers B，Beckel A，Krieger A，et al.，2013a. Shape-memory natural rubber：an exceptional material for strain and energy storage[J]. Macromolecular Chemistry and Physics，214（8）：912–923.

Heuwers B，Quitmann D，Hoeher R，et al.，2013b. Stress-induced stabilization of crystals in shape memory natural rubber[J]. Macromolecular Rapid Communications，34（2）：180–184.

Hillebrand A，Post J J，Wurbs D，et al.，2012. Down-regulation of small rubber particle protein expression affects integrity of rubber particles and rubber content in *Taraxacum brevicorniculatum*[J]. PLoS One，7（7）：e41874.

Huneau B，2011. Strain-induced crystallization of natural rubber：a review of X-ray diffraction investigations[J]. Rubber Chemistry and Technology，84（3）：425–452.

Imbernon L，Norvez S，2016a. From landfilling to vitrimer chemistry in rubber life cycle[J]. European Polymer Journal，82：347–376.

Imbernon L，Norvez S，Leibler L，2016b. Stress relaxation and self-adhesion of rubbers with exchangeable links[J]. Macromolecules，49（6）：2 172–2 178.

International Rubber Research and Development Board，2006. Portrait of the global rubber industry[M]. Kuala Lumpur：IRRDB，73–86.

Jaratrotkamjorn R，Khaokong C，Tanrattanakul V，2012. Toughness enhancement of poly（lactic acid）by melt blending with natural rubber[J]. Journal of Applied Polymer Science，124（6）：5 027–5 036.

Karger-Kocsis J，Mészáros L，Bárány T，2013. Ground tyre rubber（GTR）in thermoplastics，thermosets，and rubbers[J]. Journal of Materials Science，48（1）：1–38.

Li D J，Deng Z，Qin B，et al.，2012. De novo assembly and characterization of bark transcriptome using Illumina sequencing and development of EST-SSR markers in rubber tree（*Hevea brasiliensis* Muell. Arg.）[J]. BMC Genomics，13（1）：192.

Li Z，Fox J M，2012. Mapping rubber tree growth in mainland Southeast Asia using time-series MODIS 250m NDVI and statistical data[J]. Applied Geography，32（2）：420–432.

Morris S A，Yen G G，Wu Z，et al.，2003. Time line visualization of research fronts[J]. Journal of the Association for Information Science and Technology，54（5）：413–422.

Mondal M，Gohs U，Wagenknecht U，et al.，2013. Additive free thermoplastic vulcanizates based on natural rubber[J]. Materials Chemistry and Physics，143（1）：360–366.

Montoro P，Wu S，Favreau B，et al.，2018. Transcriptome analysis in *Hevea brasiliensis* latex revealed changes in hormone signalling pathways during ethephon stimulation and consequent Tapping Panel Dryness[J]. Scientific Reports，8（1）：1–12.

Myhre M，Saiwari S，Dierkes W，et al.，2012. Rubber recycling：chemistry，processing，and applications[J]. Rubber Chemistry and Technology，85（3）：408–449.

Nawamawat K，Sakdapipanich J T，Ho C C，et al.，2011. Surface nanostructure of *Hevea brasiliensis* natural rubber latex particles[J]. Colloids and Surfaces A：Physicochemical and Engineering Aspects，390（1-3）：157–166.

Papageorgiou D G，Kinloch I A，Young R J，2015. Graphene/elastomer nanocomposites[J]. Carbon，95：460–484.

Persson O，1994. The intellectual base and research fronts of JASIS 1986–1990[J]. Journal of the Association for Information Science and Technology，45（1）：31–38.

Price D J，1965. Networks of scientific papers[J]. Science，149（3 683）：510–515.

Potts J R，Shankar O，Du L，et al.，2012. Processing–morphology–property relationships and composite theory analysis of reduced graphene oxide/natural rubber nanocomposites[J]. Macromolecules，45（15）：6 045–6 055.

Potts J R，Shankar O，Murali S，et al.，2013. Latex and two-roll mill processing of thermally-exfoliated graphite oxide/natural rubber nanocomposites[J]. Composites Science and Technology，74：166–172.

Pérez-Aparicio R，Schiewek M，Valentín J L，et al.，2013a. Local chain deformation and overstrain in reinforced elastomers：an NMR study[J]. Macromolecules，46（14）：5 549–5 560.

Pérez-Aparicio R，Vieyres A，Albouy P A，et al.，2013b. Reinforcement in natural rubber elastomer nanocomposites：breakdown of entropic elasticity[J]. Macromolecules，46（22）：8 964–8 972.

Quitmann D，Gushterov N，Sadowski G，et al.，2013. Solvent-sensitive reversible stress-response of shape memory natural rubber[J]. ACS Applied Materials & Interfaces，5（9）：3 504–3 507.

Rahman A Y A，Usharraj A O，Misra B B，et al.，2013. Draft genome sequence of the rubber tree *Hevea brasiliensis*[J]. BMC Genomics，14（1）：75.

Riyajan S A，Sasithornsonti Y，Phinyocheep ，2012. Green natural rubber-g-modified starch for controlling urea release[J]. Carbohydrate Polymers，89（1）：251–258.

Rublon P，Huneau B，Saintier N，et al.，2013. In situ synchrotron wide-angle X-ray diffraction investigation of fatigue cracks in natural rubber[J]. Journal of Synchrotron Radiation，20（1）：105–109.

Sansatsadeekul J，Sakdapipanich J，Rojruthai P，2011. Characterization of associated proteins and phospholipids in natural rubber latex[J]. Journal of Bioscience and Bioengineering，111（6）：628–634.

Shi J W，Jiang K，Ren D Y，et al.，2013. Structure and performance of reclaimed rubber obtained by different methods[J]. Journal of Applied Polymer Science，129（3）：999–1 007.

Shneider A M，2009. Four stages of a scientific discipline；four types of scientist[J]. Trends in Biochemical Sciences，34（5）：217–223.

Small H，Griffith B C，1974. The structure of scientific literatures I：identifying and graphing specialties[J]. Social Studies of Science，4（1）：17–40.

Srivastava S，Mishra Y，2018. Nanocarbon reinforced rubber nanocomposites：detailed insights about mechanical，dynamical mechanical properties，payne，and mullin effects[J]. Nanomaterials，8（11）：945.

Tang C R，Huang D B，Yang J H，et al.，2010. The sucrose transporter *HbSUT3* plays an active role in sucrose loading to laticifer and rubber productivity in exploited trees of *Hevea brasiliensis*（para rubber tree）[J]. Plant，Cell and Environment，33（10）：1 708–1 720.

Tang C R，Yang M，Fang Y J，et al.，2016. The rubber tree genome reveals new insights into rubber production and species adaptation[J]. Nature Plants，2（6）：16 073.

Toki S，Che J，Rong L，et al.，2013. Entanglements and networks to strain-induced crystallization and stress–strain relations in natural rubber and synthetic polyisoprene at various temperatures[J]. Macromolecules，46（13）：5 238–5 248.

Triwitayakorn K，Chatkulkawin P，Kanjanawattanawong S，et al.，2011. Transcriptome sequencing of *Hevea brasiliensis* for development of microsatellite markers and construction of a genetic linkage map[J]. DNA Research，18（6）：471–482.

Vieyres A，Pérez-Aparicio R，Albouy P A，et al.，2013. Sulfur-cured natural rubber elastomer networks：correlating cross-link density，chain orientation，and mechanical response by combined techniques[J]. Macromolecules，46（3）：889–899.

Warren-Thomas E，Dolman P M，Edwards D P，2015. Increasing demand for natural rubber necessitates a robust sustainability initiative to mitigate impacts on tropical biodiversity[J]. Conservation Letters，8（4）：230–241.

White H D，McCain K W，1998. Visualizing a discipline：an author co-citation analysis of information science，1972-1995[J]. Journal of the American Society for Information Science，49（4）：327–355.

Wu J R，Xing W，Huang G S，et al.，2013. Vulcanization kinetics of graphene/natural rubber nanocomposites[J]. Polymer，54（13）：3 314–3 323.

Wu S W，Tang Z H，Guo B C，et al.，2013. Effects of interfacial interaction on chain dynamics of rubber/graphene oxide hybrids：a dielectric relaxation spectroscopy study[J]. RSC Advances，3（34）：14 549–14 559.

Xiang H P，Rong M Z，Zhang M Q，2016. Self-healing，reshaping，and recycling of vulcanized chloroprene rubber：a case study of multitask cyclic utilization of cross-linked polymer[J]. ACS Sustainable Chemistry & Engineering，4（5）：2 715–2 724.

Xu C H，Cao L M，Lin B F，et al.，2016a. Design of self-healing supramolecular rubbers by introducing ionic cross-links into natural rubber via a controlled vulcanization[J]. ACS Applied Materials & Interfaces，8（27）：17 728–17 737.

Xu C H，Huang X H，Li C H，et al，2016b. Design of "Zn^{2+} Salt-Bondings" cross-linked carboxylated styrene butadiene rubber with reprocessing and recycling ability via rearrangements of ionic cross-linkings[J]. ACS Sustainable Chemistry & Engineering，4（12）：6 981–6 990.

Xu J C，Grumbine R E，Beckschäfer P，2014. Landscape transformation through the use of ecological and socioeconomic indicators in Xishuangbanna，Southwest China，Mekong Region[J]. Ecological Indicators，36：749–756.

Zhan Y H，Lavorgna M，Buonocore G，et al.，2012. Enhancing electrical conductivity of rubber composites by constructing interconnected network of self-assembled graphene with latex mixing[J]. Journal of Materials Chemistry，22（21）：10 464–10 468.

Zhan Y H，Wu J K，Xia H S，et al.，2011. Dispersion and exfoliation of graphene in rubber by an ultrasonically-assisted latex mixing and in situ reduction process[J]. Macromolecular Materials and Engineering，296（7）：590–602.

Zhang C M，Wang W W，Huang Y，et al.，2013. Thermal，mechanical and rheological properties of polylactide toughened by expoxidized natural rubber[J]. Materials & Design，45：198–205.

Zou Z，Gong J，An F，et al.，2015. Genome-wide identification of rubber tree（*Hevea brasiliensis* Muell. Arg.）aquaporin genes and their response to ethephon stimulation in the laticifer，a rubber-producing tissue[J]. BMC Genomics，16（1）：1 001.

9　国际天然橡胶新兴趋势计量分析

9.1　引言

天然橡胶是一种以顺式-1，4-聚异戊二烯为主要成分的生物高分子。在热带农业中，天然橡胶具有特殊和重要的地位。科学技术的发展是推动天然橡胶产业升级的动力，当前一些高精尖的技术领域，如天然高分子、纳米微粒补强天然橡胶合成纳米复合材料、天然橡胶微生物降解、产胶植物橡胶生物合成与调控等（国家天然橡胶产业技术体系，2016；International Rubber Research and Development Board，2006），都可能极大地推动天然橡胶产业的发展。天然橡胶研究涉及高分子材料科学、工程化学、植物科学、生物化学与分子生物学等多个学科领域。随着这些学科领域的快速发展，紧跟集体知识发展的新趋势和关键转折至关重要。

新兴研究主题（新兴趋势）是当下新出现且呈现增长趋势的研究主题，是在当下和未来具有发展潜力的一类主题，包括对“研究前沿”的探索，因此与“研究前沿”存在交叉，并且在时间上具有新颖性，呈现“年轻化”和“快增长性”的趋势（Rotolo et al.，2015）。科学计量学是信息科学的一个分支，它通过定量分析科学文献中的模式，帮助我们识别研究领域的知识结构和新兴趋势（Chen，2017）。文献的可视化分析提供了一种有价值的、及时的、可重复和灵活的方法，除了用于传统的文献综述之外，还可以用于跟踪新趋势的发展并识别关键证据（Chen et al.，2014；Chen，2017；Chen and Leydesdorff，2014；陈超美，2015）。一系列知识图谱工具广泛用于科学计量研究，如HistCite、VOSviewer、Network WorkBench和

CiteSpace等（Garfield，2004；van Eck and Waltman，2009；Börner et al.，2010；Chen et al.，2014；Chen，2017；Chen and Leydesdorff，2014；陈超美，2015）。CiteSpace是累加式知识域分析工具，用于对科学领域中文献的新兴模式和重要变革进行可视化分析（Chen et al.，2014；Chen，2017；Chen and Leydesdorff，2014；陈超美，2015）。本章采用科学计量方法，通过CiteSpace技术对2004—2018年Web of Science核心合集的天然橡胶文献进行共被引、聚类和突现性分析，揭示国际天然橡胶领域的知识结构和新兴研究主题，为产业发展和升级提供科学参考依据。

9.2 数据来源与研究方法

9.2.1 数据来源

数据来源于Web of Science核心合集的Science Citation Index Expanded（SCI-E）和Social Sciences Citation Index（SSCI）。本书定义的天然橡胶，是指从巴西橡胶树（*Hevea brasiliensis*）、银色橡胶菊（*Parthenium argentatum*）、蒲公英属橡胶草（*Taraxacum brevicorniculatum*）等提取的天然胶乳。构建检索式TS=[“rubber tree*” or “*Hevea*” or “natur* rubber” or “natur* latex” or “nr latex” or “rubber latex” or “lactiferous plant” or “rubber yielding plant” or “rubber producing plant*” or “rubber producing crop*”] or [（“rubber”）and（“*Parthenium argentatum*” or “guayule” or “*Taraxacum brevicorniculatum*” or “dandelion” or “taraxacum” or “koksaghyz” or “rubber dandelion” or “gutta” or “tapping” or “pla-ntation*” or “yard” or “yield” or “productivity” or “garden” or “forest” or “NR” or “planting area*” or “planting” or “growing area*” or “growing state” or “intercrop*” or “interplanting” or “latex production” or “latex drainage” or “eucommia ulmoides gum”）]进行检索。检索时间范围为2004—2018年，选择文献类型“Article”或“Review”，选择语种“English”。经检索共得到9 915篇原始文献，除重后获得9 789篇文献，包含9 403篇研究性论文、386篇综述及209 528篇引文（检索日期：2019-01-21，数据库更新日期：2019-01-20）。

9.2.2 研究方法

利用CiteSpace（5.4.R1）对2004—2018年Web of Science收录的天然橡胶文献进行共引网络的生成和分析。按照每一年一个时间片段的划分方法，将15年的时间段划分为15个一年的时间片段。选取每1年中被引频次排名前100的高被引文章，构建当年时间片段的共引网络。将一系列时间片段网络合并成一个综合网络，合并后的网络包含每一个独立网络中的所有节点。节点采用由引证环充填的圆圈表示，节点的大小与标准化之后的引用次数成比例，每一个节点标签的大小和文献引用次数成比例。一个节点的特殊性采用红色和紫色两个颜色来标识。红色表示一篇引文的突现性在相应的时间片段内被探测出来。如果一个节点的中介中心性值大于0.1，节点将被加上一个紫色环，环的厚度与节点的中心性值成比例（Chen et al.，2014；Chen，2017；Chen and Leydesdorff，2014；陈超美，2015）。两个节点之间的连线表示它们的共被引关系，连线的粗细与共被引强度成比例。连线的颜色表示在该网络中连线最早出现的时间。

9.3 天然橡胶文献所属期刊的学科分布

通过对期刊的聚类，能有效地表征科学领域的分布。期刊聚类图谱共包含左右两个图，施引期刊信息叠加到左边的图上，被引期刊信息叠加到右边的图上，左右期刊之间的引证连线完整地展示引用的来龙去脉。能够快速识别所关注的科学知识文献所发表的期刊、所引用的期刊以及所发表期刊与所引用期刊之间的引证关系（Chen and Leydesdorff，2014）。期刊双图叠加显示（图9-1），左侧是施引文献所在的期刊分布，代表了天然橡胶研究所属的主要学科；右侧是对应被引文献所在的期刊分布，代表了天然橡胶研究主要引用了哪些学科的文献，并分别以紫色、蓝色和黄色等曲线表示。例如，在物理、材料和

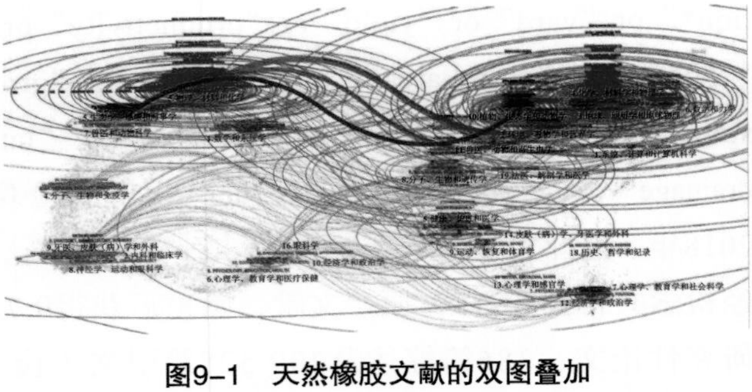

图9-1 天然橡胶文献的双图叠加

Fig. 9-1 A dual-map overlay of the natural rubber literature

化学学科，生态学、地球和海事学以及分子、生物和免疫学学科都发表了天然橡胶领域的论文，其中物理、材料和化学学科建立在右侧两个学科（化学、材料学和物理学，环境、毒物学和营养学）的基础上。

9.4 天然橡胶领域的知识结构

CiteSpace将共被引网络划分为多个共被引文献簇，使得同一簇内的文献紧密相连，而不同簇间的文献松散相连（图9-2）。表9-1按照每个聚类中成员数量的大小列出了前7个主要聚类。成员较少的聚类往往不如较大的聚类具有代表性，因为较小的聚类可能由少量文献的引用行为构成。聚类的质量也反映在轮廓值上，轮廓值是衡量聚类同质性或一致性的指标（Garfield，2004）。同质聚类的平均轮廓值趋于1。表9-1中的大多数聚类都是高度同质的。每个聚类标签是从施引文献的标题、摘要或关键词中提取。聚类的平均出版年表示其最近引用文献的时间。例如，聚类#1应变诱导结晶的平均年份是2007年。最近形成的聚类有#2、#4和#7，平均年份分别是2009年、2010年和2011年。在后续的讨论中，选择聚类中被引次数最多的文献和覆盖率最高的施引文献进行分析。

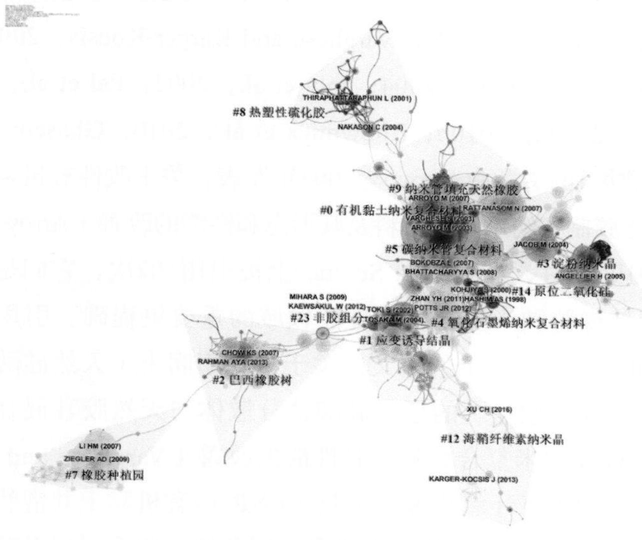

图9-2 天然橡胶文献共被引网络

Fig. 9-2 Natural rubber research shown in a network of co-cited references

表9-1 共被引文献的聚类

Tab. 9-1 Clusters of co-cited references

聚类编号 Cluster ID	大小 Size	轮廓值 Silhouette	平均年 Mean year	聚类命名 Cluster label
0	89	0.863	2003	有机黏土纳米复合材料
1	81	0.935	2007	应变诱导结晶
2	72	0.990	2009	巴西橡胶树
3	60	0.972	2005	淀粉纳米晶
4	58	0.926	2011	氧化石墨烯纳米复合材料
5	58	0.835	2008	碳纳米管复合材料
7	49	0.986	2011	橡胶种植园
8	46	0.997	2002	热塑性硫化胶
9	45	0.964	2005	纳米管填充天然橡胶
12	28	0.98	2012	海鞘纤维素纳米晶
14	21	0.984	2001	原位二氧化硅
23	4	0.995	2008	非胶组分

9.4.1 聚类#0——有机黏土纳米复合材料

聚类#0代表了黏土/天然橡胶纳米复合材料研究重要的施引文献和引文（表9-2）（Arroyo et al.，2003；Varghese and Karger-Kocsis，2003；Ray and Okamoto，2003；Teh et al.，2004；Joly et al.，2002；Pal et al.，2010；Chakraborty et al.，2010a，2010b；Bhowmick et al.，2010；Ghasemi et al.，2010）。引用次数第1的文章由Arroyo等2003年发表，关于改性有机蒙脱土替代炭黑填充天然橡胶制备纳米复合材料及其形态和性能的改善（Arroyo et al.，2003）。截至2019年3月，在Web of Science上被引用502次，它们是层状硅酸盐晶体增强天然橡胶纳米复合材料性能领域的重要里程碑。引用次数第2的是Varghese和Karger-Kocsis的论文，关于钠基膨润土（天然硅酸盐）和钠氟锂蒙脱石（合成硅酸盐）的层状硅酸盐分散体与天然胶乳混合，制备层状硅酸盐/天然橡胶纳米复合材料及其性能的改善（Varghese and Karger-Kocsis，2003）。Pal等将环氧化天然橡胶（ENR）/有机黏土共混物与天然橡胶（NR）/丁苯橡胶（SBR）共混物混合，制备纳米复合材料及其对形态结构、力学性能和热稳定性等的影响（Pal et al.，2010），Chakraborty等研究钠—改性有机膨润土/SBR纳米复合材料的物理性能（Chakraborty et al.，2010a），也在一定程度上揭示了与主题的一致性。

表9-2 聚类#0有机黏土纳米复合材料的被引文献和施引文献

Tab. 9-2 Cited references and citing articles of cluster #0 organoclay nanocomposite

被引文献Cited references		施引文献Citing articles	
引用次数 Cites	作者（年）期刊，卷，页 Author（Year）Journal，Volume，Page	覆盖率（%）Coverage（%）	作者（年）标题 Author（Year）Title
182	Arroyo M（2003）Polymer，V44，P2447	24	Pal，Kaushik（2010）Influence of fillers on nr/sbr blends containing enr-organoclay nanocomposites：morphology and wear
94	Varghese S（2003）Polymer，V44，P4921	19	Chakraborty，Sugata（2010a）Effect of treatment of bis（3-triethoxysilyl propyl）tetrasulfane on physical property of in situ sodium activated and organomodified bentonite clay-sbr rubber nanocomposite
91	Ray S S（2003）Prog Polym Sci，V28，P1539	16	Chakraborty，Sugata（2010b）Study of the properties of in-situ sodium activated and organomodified bentonite clay-sbr rubber nanocomposites-part i：characterization and rheometric properties
87	Teh P L（2004）Eur Polym J，V40，P2513	14	Bhowmick，Anil K（2010）Exfoliation of nanolayer assemblies for improved natural rubber properties：methods and theory
80	Joly S（2002）Chem Mater，V14，P4202	13	Ghasemi I（2010）Evaluating the effect of processing conditions and organoclay content on the properties of styrene-butadiene rubber/organoclay nanocomposites by response surface methodology

9.4.2 聚类#1——应变诱导结晶

聚类#1的核心成员进一步揭示了天然橡胶纳米复合材料的应变诱导结晶行为和分子取向（表9-3）（Tosaka et al.，2004；Toki et al.，2009；Diani et al.，2009；Carretero-González et al.，2008；Huneau，2011；Weng et al.，2010；Chen et al.，2012；Kimura et al.，2010；Xu et al.，2012a，2012b）。尤其是Tosaka等采用同步X射线衍射法研究不同网链密度的天然橡胶材料应变诱导结晶行为（Tosaka et al.，2004）；Toki等通过原位同步广角X射线衍射法研究硫化天然橡胶在单轴变形过程中的分子取向和应变诱导结晶行为（Toki et al.，2009）。施引文献中（Weng et al.，2010；Chen et al.，2012；Kimura et al.，2010；Xu et al.，2012a，2012b），Weng等关于硫化拉伸应变下，原位同步X射线衍射技术对充填和未充填多壁碳纳米管天然橡胶诱导结晶行为（Weng et al.，2010），Chen等关于原位聚合甲基丙烯酸锌（ZDMA）/天然橡胶纳米复合材料的网络演化和黏弹性行为（Chen et al.，2012）。

表9-3　聚类#1应变诱导结晶的被引文献和施引文献

Tab. 9-3　Cited references and citing articles of cluster #1 strain-induced crystallization

被引文献Cited references		施引文献Citing articles	
引用次数 Cites	作者（年）期刊，卷，页 Author（Year）Journal，Volume，Page	引用次数 Cites	作者（年）期刊，卷，页 Author（Year）Journal，Volume，Page
74	Tosaka M（2004）Macromolecules，V37，P3299	10	Weng G S（2010）Large-scale orientation in a vulcanized stretched natural rubber network: proved by in situ synchrotron x-ray diffraction characterization
72	Toki S（2002）Macromolecules，V35，P6578	7	Chen Y K（2012）Viscoelasticity behaviors of lightly cured natural rubber/zinc dimethacrylate composites
57	Diani J（2009）Eur Polym J，V45，P601	7	Kimura，Hideaki（2010）Molecular dynamics and orientation of stretched rubber by solid-state c-13 nmr
55	Carretero-gonzalez J（2008）Macromolecules，V4，P6763	7	Xu C H（2012b）A study on the crosslink network evolution of magnesium dimethacrylate/natural rubber composite
54	Huneau B（2011）Rubber Chem Technol，V84，P425	7	Xu C H（2012a）Thermal aging on mechanical properties and crosslinked network of natural rubber/zinc dimethacrylate composites

9.4.3　聚类#2——巴西橡胶树

聚类#2是较活跃的聚类（表9-4），该聚类的被引文献中（Chow et al.，

2007；Rahman et al.，2013；van Beilen et al.，2007；Ko et al.，2003；Li et al.，2012），引用次数最高的是Chow等2007年的文章，通过转录组分析揭示橡胶胶乳与橡胶生物合成、应激或防御反应相关的重要编码蛋白（Chow et al.，2007）。引用次数第2的是Rahman等关于构建橡胶树基因组序列框架图，鉴定了与橡胶生物合成、橡胶木材形成、抗病性和致敏性相关的关键基因（Rahman et al.，2013）。施引文献中（Berthelot et al.，2014；Wei et al.，2015；Nie et al.，2016；Duan et al.，2013；Lopez et al.，2016），Berthelot等综述了与橡胶树橡胶粒子相关的两种重要蛋白的进展：橡胶延伸因子（REF）和小橡胶粒子蛋白（SRPP）（Berthelot et al.，2014）。Lopez等在巴西橡胶树基因组中发现3个XIP基因（*HbXIP1; 1*、*HbXIP2; 1*和*HbXIP3; 1*），特别是*HbXIP2; 1*与胶乳代谢的分子和生理反应密切相关（Lopez et al.，2016）。

表9-4　聚类#2巴西橡胶树的被引文献和施引文献

Tab. 9-4　Cited references and citing articles of cluster #2 *Hevea brasiliensis*

被引文献Cited references		施引文献Citing articles	
引用次数 Cites	作者（年）期刊，卷，页 Author（Year）Journal，Volume，Page	引用次数 Cites	作者（年）期刊，卷，页 Author（Year）Journal，Volume，Page
81	Chow K S（2007）J Exp Bo，V58，P2429	8	Berthelot，Karine（2014）*Hevea brasiliensis* ref（hev b1）and srpp（hev b3）：an overview on rubber particle proteins
65	Rahman A Y A（2013）Bmc Genomics，V14，P75	7	Wei F（2015）Transcriptome sequencing and comparative analysis reveal long-term flowing mechanisms in *hevea brasiliensis* latex
64	van Beilen J（2007）Trends Biotechnol，V25，P522	7	Nie Z Y（2016）Profiling ethylene-responsive genes expressed in the latex of the mature virgin rubber trees using cdna microarray
59	Ko J H（2003）Plant Mol Biol，V53，P479	7	Duan C F（2013）Identification of the *Hevea brasiliensis* ap2/erf superfamily by rna sequencing
50	Li D J（2012）Bmc Genomics，V13，P192	7	Lopez，David（2016）The *Hevea brasiliensis* xip aquaporin subfamily：genomic，structural and functional characterizations with relevance to intensive latex harvesting
47	Tang C R（2010）Plant Cell Environ，V33，P1708	7	Rahman，Ahmad Yamin Abdul（2013）Draft genome sequence of the rubber tree *Hevea brasiliensis*

（续表）

被引文献Cited references		施引文献Citing articles	
引用次数 Cites	作者（年）期刊，卷，页 Author（Year）Journal，Volume，Page	引用次数 Cites	作者（年）期刊，卷，页 Author（Year）Journal，Volume，Page
44	Tang C R（2016）Nat Plants，V2，P16073	7	Wang D（2016）A protein extraction method for low protein concentration solutions compatible with the proteomic analysis of rubber particles
41	Triwitayakorn K（2011）Dna Res，V18，P471	6	Berthelot，Karine（2014）Homologous *Hevea brasiliensis* ref（hev b1）and srpp（hev b3）present different auto-assembling
40	Chow K S（2012）J Exp Bot，V63，P1863	6	Cornish，Katrina（2018）Unusual subunits are directly involved in binding substrates for natural rubber biosynthesis in multiple plant species
37	Nawamawat K（2011）Colloid Surface A，V390，P157	6	Liu J P（2016）Molecular mechanism of ethylene stimulation of latex yield in rubber tree（*Hevea brasiliensis*）revealed by de novo sequencing and transcriptome analysis

9.4.4 聚类#3——淀粉纳米晶

聚类#3代表了淀粉、纤维素等植物多糖提取的微晶纤维素增强橡胶纳米复合材料性能研究（表9-5）（Jacob et al.，2004；Angellier et al.，2005a，2005b；Abdelmouleh et al.，2007；Gopalan Nair and Dufresne，2003；Dufresne，2010；Pasquini et al.，2010；Sriupayo et al.，2005a，2005b；Zeng et al.，2010）。引用次数最高的文章是Jacob等2004年发表，油棕、剑麻混合植物纤维增强天然橡胶纳米复合材料机械性能（Jacob et al.，2004）。引用次数第2的是Angellier等2005年发表，以糯玉米淀粉纳米晶悬浮液为填料与天然胶乳混合制备纳米复合材料，从形态结构和阻隔性能等对其表征（Angellier et al.，2005a）。覆盖率较高的5篇文献中引用了该聚类11%～20%的引文（Dufresne，2010；Pasquini et al.，2010；Sriupayo et al.，2005a，2005b；Zeng et al.，2010）。Dufresne综述了以纤维素、甲壳素或淀粉为原料，经酸水解制备多糖（纤维素、甲壳素或淀粉）纳米晶悬浮液增强聚合物纳米复合材料性能（Dufresne，2010）。Pasquini等将木薯、甘蔗渣作为提取纤维素晶须的原料用于增强天然橡胶纳米复合材料性能（Pasquini et al.，2010）。

表9-5 聚类#3淀粉纳米晶的被引文献和施引文献

Tab. 9-5 Cited references and citing articles of cluster #3 starch nanocrystal

被引文献Cited references		施引文献Citing articles	
引用次数 Cites	作者（年）期刊，卷，页 Author（Year）Journal，Volume，Page	引用次数 Cites	作者（年）期刊，卷，页 Author（Year）Journal，Volume，Page
79	Jacob M（2004）Compos Sci Technol，V64，P955	20	Dufresne，Alain（2010）Processing of polymer nanocomposites reinforced with polysaccharide nanocrystals
63	Angellier H（2005）Macromolecules，V38，P9161	18	Pasquini，Daniel（2010）Extraction of cellulose whiskers from cassava bagasse and their applications as reinforcing agent in natural rubber
59	Angellier H（2005）Macromolecules，V38，P3783	12	Sriupayo J（2005b）Preparation and characterization of α-chitin whisker-reinforced poly（vinyl alcohol）nanocomposite films with or without heat treatment
55	Abdelmouleh M（2007）Compos Sci Technol，V67，P1627	12	Sriupayo J（2005a）Preparation and character ization of alpha-chitin whisker-reinforced chitosan nanocomposite films with or without heat treatment
54	Nair K G（2003）Biomacromolecules，V4，P666	11	Zeng M（2010）Preparation and characterization of nanocomposite films from chitin whisker and waterborne poly（ester-urethane）with or without ultra-sonification treatment

9.4.5 聚类#4——氧化石墨烯纳米复合材料

聚类#4是最近形成的聚类（表9-6）。该聚类由58篇共被引文献组成，被引文献中（Potts et al.，2012，2013；Zhan et al.，2011；Tang et al.，2012；Hernández et al.，2012），引用次数最高的文章是Potts等2012年发表，展示了胶乳共混法制备还原氧化石墨烯/天然橡胶纳米复合材料显著提高其机械、导电和导热性能（Potts et al.，2012）。第二大引用文献展示了超声辅助胶乳共混法和原位法结合制备氧化石墨烯/天然橡胶纳米材料，提高基体拉伸强度（Zhan et al.，2011）。入选的5篇施引文献引用了该聚类9%～13%的引文。引用覆盖率较高的是Mensah等、Wu等和Papageorgiou等的文章，占13%。主要关注的是不同类型石墨烯补强天然橡胶或弹性体纳米复合材料及其对材料结构、导电、导热、屏蔽、机械、力学和硫化性能等的影响（Mensah et al.，2018；Wu et al.，2013；Papageorgiou et al.，2015）。其他施引文献也在关注改进的方法制备石墨烯或其衍生物（氧化石墨烯）/天然橡胶纳米复合材料及其性能的增强（Wu et al.，2013；Srivastava and Mishra，2018）。

表9-6　聚类#4氧化石墨烯纳米复合材料的被引文献和施引文献

Tab. 9-6　Cited references and citing articles of cluster #4 graphene oxide nanocomposite

被引文献Cited references		施引文献Citing articles	
引用次数 Cites	作者（年）期刊，卷，页 Author（Year）Journal，Volume，Page	引用次数 Cites	作者（年）期刊，卷，页 Author（Year）Journal，Volume，Page
91	Potts J R（2012）Macromolecules，V45，P6045	13	Mensah，Bismark（2018）Graphene-reinforced elastomeric nanocomposites：a review
70	Zhan Y H（2011）Macromol Mater Eng，V296，P590	13	Wu S W（2013）Effects of interfacial interaction on chain dynamics of rubber/graphene oxide hybrids：a dielectric relaxation spectroscopy study
56	Zhan Y H（2012）J Mater Chem，V22，P10464	13	Papageorgiou，Dimitrios G（2015）Graphene/elastomer nanocomposites
51	Potts J R（2013）Compos Sci Technol，V74，P166	11	Wu J R（2013）Vulcanization kinetics of graphene/natural rubber nanocomposites
48	Hernandez M（2012）Compos Sci Technol，V73，P40	9	Srivastava，Suneel Kumar（2018）Nanocarbon reinforced rubber nanocomposites：detailed insights about mechanical，dynamical mechanical properties，payne，and mullin effects
43	Wu J R（2013）Polymer，V54，P3314	9	Berki，Peter（2017）Natural rubber/graphene oxide nanocomposites via melt and latex compounding：comparison at very low graphene oxide content
38	Tang Z H（2012）J Mater Chem，V22，P7492	9	Li C P（2013）Ammonium-assisted green fabrication of graphene/natural rubber latex composite

（续表）

被引文献Cited references		施引文献Citing articles	
引用次数 Cites	作者（年）期刊，卷，页 Author（Year）Journal，Volume，Page	引用次数 Cites	作者（年）期刊，卷，页 Author（Year）Journal，Volume，Page
37	Wu J R（2013）Polymer，V54，P1930	9	Liu X A（2015）Preparation of rubber/graphene oxide composites with in-situ interfacial design
36	Kim H（2010a）Macromolecules，V43，P6515	9	Mondal，Titash（2017）Graphene-based elastomer nanocomposites：functionalization techniques，morphology，and physical properties
33	Ozbas B（2012）J Polym Sci Pol Phys，V50，P718	9	Luo Y Y（2014）Fabrication of conductive elastic nanocomposites via framing intact interconnected graphene networks

9.4.6 聚类#5——碳纳米管复合材料

聚类#5也是持续发展的，代表了碳纳米管、多壁碳纳米管等填充天然橡胶或弹性体制备复合材料的相关研究（表9-7）（Bokobza，2007，2012；Bhattacharyya et al.，2008；Fröhlich et al.，2005；Shanmugharaj et al.，2007；Das et al.，2008；Peng et al.，2010；Kapgate et al.，2014；Sittiphan et al.，2014；Xu et al.，2017）。引用次数最高的论文综述了碳纳米管作为弹性体增强填料的内在潜力（Bokobza，2007），引用次数第2高的论文是多壁碳纳米管分散于天然橡胶增强其拉伸强度和动力学性能（Bhattacharyya et al.，2008）。聚类中覆盖率较高的施引文献（Bokobza，2012；Peng et al.，2010；Kapgate et al.，2014；Sittiphan et al.，2014；Xu et al.，2017），有超声法制备多壁碳纳米管填充天然橡胶复合材料及其制备工艺条件对材料导电性能的影响（Bokobza，2012）；采用乳液共混法制备多壁碳纳米管/天然橡胶复合材料及碳纳米管与胶乳粒子之间的相互作用机制（Peng et al.，2010）。

表9-7　聚类#5碳纳米管复合材料的被引文献和施引文献

Tab. 9-7　Cited references and citing articles of cluster #5 carbon nanotube composite

被引文献Cited references		施引文献Citing articles	
引用次数 Cites	作者（年）期刊，卷，页 Author（Year）Journal，Volume，Page	引用次数 Cites	作者（年）期刊，卷，页 Author（Year）Journal，Volume，Page
93	Bokobza L（2007）Polymer，V48，P4907	10	Bokobza，Liliane（2012）Enhanced electrical and mechanical properties of multiwall carbon nanotube rubber composites
91	Bhattacharyya S（2008）Carbon，V46，P1037	7	Kapgate，Bharat P（2014）Effect of silane integrated sol-gel derived in situ silica on the properties of nitrile rubber
86	Frohlich J（2005）Compos Part A-Appl S，V36，P449	6	Sittiphan，Torpong（2014）Styrene grafted natural rubber reinforced by in situ silica generated via sol-gel technique
82	Shanmugharaj A M（2007）Compos Sci Technol，V67，P1813	6	Xu T W（2017）Effect of acetone extract from natural rubber on the structure and interface interaction in nr/cb composites
67	Das A（2008）Polymer，V49，P5276	6	Peng Z（2010）Self-assembled natural rubber/multi-walled carbon nanotube composites using latex compounding techniques

9.4.7　聚类#7——橡胶种植园

聚类#7也是最近形成的聚类，代表了土地利用变化下扩张种植的橡胶林对生态环境效应的影响（表9-8）（Ziegler et al.，2009；Li et al.，2007，2008；Li et al.，2012；Qiu，2009；Wigboldus et al.，2017；Golbon and Cotter，2018；Clough et al.，2016；Guillaume et al.，2016a，2016b；Yi et al.，2014）。引用次数最高的是Ziegler等2009年发表在Science上的一篇论文，关于东南亚地区迅速扩张的橡胶种植园，在很大程度上替代现有土地的常绿阔叶林和次生植被等，这种土地利用转换对今后环境将造成的巨大影响（Ziegler et al.，2009）。引用次数第2的是Li等利用Landsat影像量化西双版纳土地利用/土地覆盖的变化，发现最明显的变化是森林覆盖的减少和橡胶林的增加（Li et al.，2007）。施引文献中（Wigboldus et al.，2017；Golbon and Cotter，2018；Clough et al.，2016；Guillaume et al.，2016a，2016b；Yi et al.，2014），Wigboldus等研究土地利用变化下云南西双版纳橡胶种植的可持续发展模式（Wigboldus et al.，2017）。Golbon等基于降水量、温度和全球环流模型等气象数据，预测气候变化对大湄公河次地区（GMS）橡胶种植园的潜在影响（Golbon and Cotter，2018）。

表9–8 聚类#7橡胶种植园的被引文献和施引文献

Tab. 9–8 Cited references and citing articles of cluster #7 rubber plantation

被引文献Cited references		施引文献Citing articles	
引用次数 Cites	作者（年）期刊，卷，页 Author（Year）Journal，Volume，Page	引用次数 Cites	作者（年）期刊，卷，页 Author（Year）Journal，Volume，Page
109	Ziegler A D（2009）Science，V324，P1024	12	Wigboldus，Seerp（2017）Scaling green rubber cultivation in southwest china-an integrative analysis of stakeholder perspectives
81	Li H M（2007）Biodivers Conserv，V16，P1731	12	Golbon，Reza（2018）Climate change impact assessment on the potential rubber cultivating area in the greater mekong subregion
59	Li Z（2012）Appl Geogr，V32，P420	10	Clough，Yann（2016）Land-use choices follow profitability at the expense of ecological functions in indonesian smallholder landscapes
58	Li H M（2008）Forest Ecol Manag，V255，P16	10	Guillaume，Thomas（2016b）Sensitivity and resistance of soil fertility indicators to land-use changes：new concept and examples from conversion of indonesian rainforest to plantations
47	Qiu J（2009）Nature，V457，P246	10	Yi Z F（2014）Can carbon-trading schemes help to protect china's most diverse forest ecosystems？A case study from Xishuangbanna，Yunnan
44	Fox J（2013）J Peasant Stud，V40，P155	10	Drescher，Jochen（2016）Ecological and socio-economic functions across tropical land use systems after rainforest conversion
40	Ahrends A（2015）Global Environ Chang，V34，P48	8	Zhang M X（2017）Natural forest at landscape scale is most important for bird conservation in rubber plantation
36	Xu J C（2014）Ecol Indic，V36，P749	8	Guillaume，Thomas（2016a）Soil degradation in oil palm and rubber plantations under land resource scarcity
33	Dong J W（2013）Remote Sens Environ，V134，P392	8	Yi Z F（2014）Developing indicators of economic value and biodiversity loss for rubber plantations in Xishuangbanna，southwest china：a case study from menglun township
33	de Blecourt M（2013）Plos One，V8，Pe69357	8	Jiang X J（2017）Land degradation controlled and mitigated by rubber-based agroforestry systems through optimizing soil physical conditions and water supply mechanisms：a case study in Xishuangbanna，china

9.5 天然橡胶领域活跃的研究主题

天然橡胶是一个跨学科的研究领域，它不仅涉及众多学科，而且从文献的学科类别、关键词及引文突变等方面体现了出版物的变化强度。在Web of Science中，每一篇文献都被指定为一个或多个学科类别，如植物科学、微生物学和材料科学等。每一篇文献还被标引了一些关键词。这些学科类别或关键词强度的迅速增加和转移表明在学科水平研究领域的活跃程度。学科类别、关键词或引文的突发性是在不同粒度级别上研究主题活跃性的有价值的指标（Chen et al.，2014；Chen，2017；Chen and Leydesdorff，2014；陈超美，2015）。

9.5.1 基于Web of Science的学科类别

突现性测度的是一个给定的频次函数在一个较小时间段内的波动显著性。对2004—2018年天然橡胶文献的学科类别进行突现性检测，发现随着时间的推移，学科类别呈现出突现强度的变化。共有144个学科类别，在44个学科类别中检测到突现性（图9-3）。蓝色线段表示时间间隔，红色线段显示一个学科出现突现性的时间段，以突现性的开始和结束年份表示持续时间（Chen et al.，2014；Chen，2017；Chen and Leydesdorff，2014；陈超美，2015）。如图9-3顶部的过敏学，在2004—2008年有一个突现增强时期，突现强度为34.816 3。2011年以前的热点学科属于过敏学、免疫学、生物技术与应用微生物。与材料科学、生物材料相关的热点学科在2011年首次出现，突现强度从2011年持续到2015年，并超过3.129 2。昆虫学、生物多样性与保护在2012年首次出现，并从2012—2015年突现性增强，突现强度超过5.028 1。2015年以来，绿色可持续科技（14.909 4）、遥感（11.153 8）和地质学（7.027 7）的突现性都非常强劲。

Subject Categories	Year	Strength	Begin	End	2004 - 2018
ALLERGY	2004	34.8163	**2004**	2008	
IMMUNOLOGY	2004	31.5111	**2004**	2008	
CHEMISTRY, ORGANIC	2004	10.9678	**2004**	2007	
BIOCHEMISTRY & MOLECULAR BIOLOGY	2004	8.1404	**2004**	2005	
DERMATOLOGY	2004	6.674	**2004**	2005	
GENERAL & INTERNAL MEDICINE	2004	4.8665	**2004**	2010	
PUBLIC, ENVIRONMENTAL & OCCUPATIONAL HEALTH	2004	4.8209	**2004**	2006	
DENTISTRY, ORAL SURGERY & MEDICINE	2004	4.6305	**2004**	2008	
SPECTROSCOPY	2004	4.4457	**2004**	2007	
MEDICINE, GENERAL & INTERNAL	2004	3.4354	**2004**	2011	
SOIL SCIENCE	2004	3.3432	**2004**	2005	
RESPIRATORY SYSTEM	2004	3.1487	**2004**	2008	
BIOCHEMICAL RESEARCH METHODS	2004	3.0796	**2004**	2007	
NUCLEAR SCIENCE & TECHNOLOGY	2004	4.6414	**2005**	2008	
PHYSICS, NUCLEAR	2004	3.4485	**2005**	2007	
BIOPHYSICS	2004	3.3444	**2005**	2007	
ZOOLOGY	2004	2.9414	**2005**	2009	
PEDIATRICS	2004	3.9396	**2006**	2008	
BIOLOGY	2004	2.961	**2007**	2008	
LIFE SCIENCES & BIOMEDICINE - OTHER TOPICS	2004	2.961	**2007**	2008	
ENGINEERING, MANUFACTURING	2004	2.9048	**2007**	2009	
BIOTECHNOLOGY & APPLIED MICROBIOLOGY	2004	5.5846	**2008**	2010	
MATERIALS SCIENCE, PAPER & WOOD	2004	3.9919	**2011**	2014	
AREA STUDIES	2004	2.8892	**2011**	2013	
ENTOMOLOGY	2004	3.9256	**2012**	2013	
ELECTROCHEMISTRY	2004	3.421	**2012**	2014	
MATERIALS SCIENCE, BIOMATERIALS	2004	3.1292	**2013**	2014	
GENETICS & HEREDITY	2004	2.8396	**2013**	2015	
HORTICULTURE	2004	5.5068	**2014**	2018	
BIODIVERSITY CONSERVATION	2004	5.0281	**2014**	2015	
BIODIVERSITY & CONSERVATION	2004	5.0281	**2014**	2015	
MULTIDISCIPLINARY SCIENCES	2004	4.3499	**2014**	2016	
ENVIRONMENTAL STUDIES	2004	3.4618	**2014**	2015	
REMOTE SENSING	2004	11.1538	**2015**	2018	
IMAGING SCIENCE & PHOTOGRAPHIC TECHNOLOGY	2004	4.2846	**2015**	2018	
PHYSICAL GEOGRAPHY	2004	4.2006	**2015**	2018	
GEOGRAPHY, PHYSICAL	2004	4.2006	**2015**	2018	
ENGINEERING, ELECTRICAL & ELECTRONIC	2004	3.6523	**2015**	2018	
GREEN & SUSTAINABLE SCIENCE & TECHNOLOGY	2004	14.9094	**2016**	2018	
GEOLOGY	2004	7.0277	**2016**	2018	
GEOSCIENCES, MULTIDISCIPLINARY	2004	6.6431	**2016**	2018	
CONSTRUCTION & BUILDING TECHNOLOGY	2004	5.6091	**2016**	2018	
SCIENCE & TECHNOLOGY - OTHER TOPICS	2004	3.1762	**2016**	2018	
TRANSPORTATION	2004	2.9341	**2016**	2018	

图9-3　2004—2018年出现突发性的学科类别

Fig. 9-3　Subject categories have occurrence bursts during 2004-2018

9.5.2　基于文献的关键词

通过探测关键词的突现模式也可以揭示研究领域的新趋势。从2004—2018年的天然橡胶文献中，每年选取出现频次前100的关键词进行突现性检测，共筛选出331个关键词。最常见的是“机械性能”，出现在1 998篇文章中，“复合材料”出现在1 248篇文献中。在突现性较强的114个关键词中（图9-4），2004—2013年，突现性较强的有“纳米黏土”（21.163 3）、“化学改性”（14.963 2）、“硅酸盐复合材料”（14.053 1）等。突现强度持续到2018年的关键词特别值得关注，如2014年开始突现出现在62篇文献中的“力学”（18.377 0）、出现在111篇文献中的“热带雨林”（15.365 2）、出现在70篇文献中的“土地利用”（12.488 9）、出现在125篇文献中的“基因表达”（7.018 1）等；2015年开始突现出现在166篇文献中的“多样性”（11.397 1）等；2016年开始突现出现在106篇文献中的“石墨烯”（30.121 5）、174篇文献中的“种植园”（15.361 3）、72篇文献中的“西双版纳”（13.618 7）、88篇文献中的“光谱学”（11.749 0）和222篇文献中的“纳米管”（8.231 7）等。

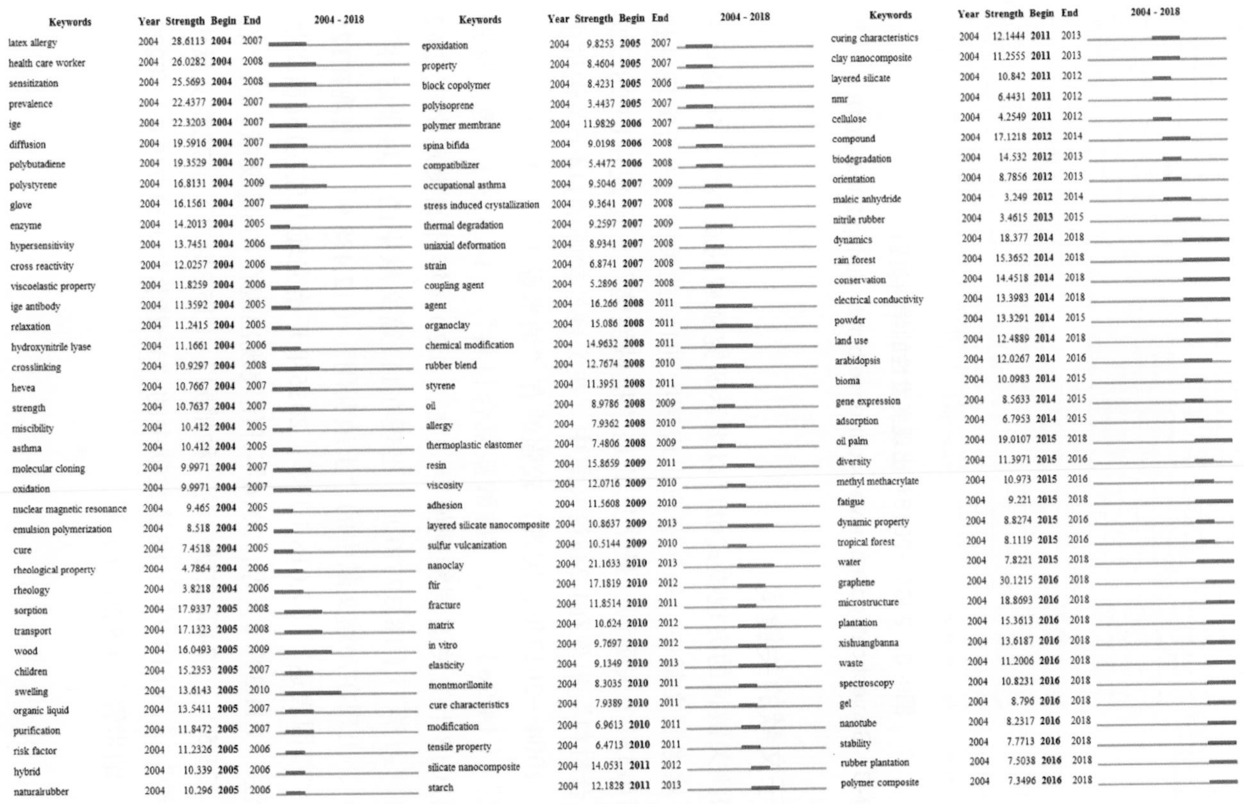
Keywords Year Strength Begin End 2004 - 2018
latex allergy 2004 28.6113 2004 2007
health care worker 2004 26.0282 2004 2008
sensitization 2004 25.5693 2004 2008
prevalence 2004 22.4377 2004 2007
ige 2004 22.3203 2004 2007
diffusion 2004 19.5916 2004 2007
polybutadiene 2004 19.3529 2004 2007
polystyrene 2004 16.8131 2004 2009
glove 2004 16.1561 2004 2007
enzyme 2004 14.2013 2004 2005
hypersensitivity 2004 13.7451 2004 2006
cross reactivity 2004 12.0257 2004 2006
viscoelastic property 2004 11.8259 2004 2006
ige antibody 2004 11.3592 2004 2005
relaxation 2004 11.2415 2004 2005
hydroxynitrile lyase 2004 11.1661 2004 2006
crosslinking 2004 10.9297 2004 2008
hevea 2004 10.7667 2004 2007
strength 2004 10.7637 2004 2007
miscibility 2004 10.412 2004 2005
asthma 2004 10.412 2004 2005
molecular cloning 2004 9.9971 2004 2007
oxidation 2004 9.9971 2004 2007
nuclear magnetic resonance 2004 9.465 2004 2005
emulsion polymerization 2004 8.518 2004 2005
cure 2004 7.4518 2004 2005
rheological property 2004 4.7864 2004 2006
rheology 2004 3.8218 2004 2006
sorption 2004 17.9337 2005 2008
transport 2004 17.1323 2005 2008
wood 2004 16.0493 2005 2009
children 2004 15.2353 2005 2007
swelling 2004 13.6143 2005 2010
organic liquid 2004 13.5411 2005 2007
purification 2004 11.8472 2005 2007
risk factor 2004 11.2326 2005 2006
hybrid 2004 10.339 2005 2006
naturalrubber 2004 10.296 2005 2006
Keywords Year Strength Begin End 2004 - 2018
epoxidation 2004 9.8253 2005 2007
property 2004 8.4604 2005 2007
block copolymer 2004 8.4231 2005 2006
polyisoprene 2004 3.4437 2005 2007
polymer membrane 2004 11.9829 2006 2007
spina bifida 2004 9.0198 2006 2008
compatibilizer 2004 5.4472 2006 2008
occupational asthma 2004 9.5046 2007 2009
stress induced crystallization 2004 9.3641 2007 2008
thermal degradation 2004 9.2597 2007 2009
uniaxial deformation 2004 8.9341 2007 2008
strain 2004 6.8741 2007 2008
coupling agent 2004 5.2896 2007 2008
agent 2004 16.266 2008 2011
organoclay 2004 15.086 2008 2011
chemical modification 2004 14.9632 2008 2011
rubber blend 2004 12.7674 2008 2010
styrene 2004 11.3951 2008 2011
oil 2004 8.9786 2008 2009
allergy 2004 7.9362 2008 2010
thermoplastic elastomer 2004 7.4806 2008 2009
resin 2004 15.8659 2009 2011
viscosity 2004 12.0716 2009 2010
adhesion 2004 11.5608 2009 2010
layered silicate nanocomposite 2004 10.8637 2009 2013
sulfur vulcanization 2004 10.5144 2009 2010
nanoclay 2004 21.1633 2010 2013
ftir 2004 17.1819 2010 2012
fracture 2004 11.8514 2010 2011
matrix 2004 10.624 2010 2012
in vitro 2004 9.7697 2010 2012
elasticity 2004 9.1349 2010 2013
montmorillonite 2004 8.3035 2010 2011
cure characteristics 2004 7.9389 2010 2011
modification 2004 6.9613 2010 2011
tensile property 2004 6.4713 2010 2011
silicate nanocomposite 2004 14.0531 2011 2012
starch 2004 12.1828 2011 2013
Keywords Year Strength Begin End 2004 - 2018
curing characteristics 2004 12.1444 2011 2013
clay nanocomposite 2004 11.2555 2011 2013
layered silicate 2004 10.842 2011 2012
nmr 2004 6.4431 2011 2012
cellulose 2004 4.2549 2011 2012
compound 2004 17.1218 2012 2014
biodegradation 2004 14.532 2012 2013
orientation 2004 8.7856 2012 2013
maleic anhydride 2004 3.249 2012 2014
nitrile rubber 2004 3.4615 2013 2015
dynamics 2004 18.377 2014 2018
rain forest 2004 15.3652 2014 2018
conservation 2004 14.4518 2014 2018
electrical conductivity 2004 13.3983 2014 2018
powder 2004 13.3291 2014 2015
land use 2004 12.4889 2014 2018
arabidopsis 2004 12.0267 2014 2016
bioma 2004 10.0983 2014 2015
gene expression 2004 8.5633 2014 2015
adsorption 2004 6.7953 2014 2015
oil palm 2004 19.0107 2015 2018
diversity 2004 11.3971 2015 2016
methyl methacrylate 2004 10.973 2015 2016
fatigue 2004 9.221 2015 2018
dynamic property 2004 8.8274 2015 2016
tropical forest 2004 8.1119 2015 2016
water 2004 7.8221 2015 2018
graphene 2004 30.1215 2016 2018
microstructure 2004 18.8693 2016 2018
plantation 2004 15.3613 2016 2018
xishuangbanna 2004 13.6187 2016 2018
waste 2004 11.2006 2016 2018
spectroscopy 2004 10.8231 2016 2018
gel 2004 8.796 2016 2018
nanotube 2004 8.2317 2016 2018
stability 2004 7.7713 2016 2018
rubber plantation 2004 7.5038 2016 2018
polymer composite 2004 7.3496 2016 2018

图9-4　2004—2018年出现突发性的关键词

Fig. 9-4　Keywords have occurrence bursts during 2004–2018

9.5.3 基于被引文献

基于每年引用次数前100的文献，探测出2004—2018年突发强度排名前100的引文（图9-5）。突发性最强的文章是Arroyo等（2003）关于层状硅酸盐纳米材料填充天然橡胶增强复合材料性能的里程碑意义的论文，从2006—2013年，突发性持续了8年，突发强度为30.147 9（Arroyo et al., 2003）。Alexandre等排名第3突发强度为23.802 1的文章（Alexandre and Dubois，2000），综述了层状硅酸盐应用于聚合物纳米复合材料的制备和性能提升，突发持续时间从2004—2010年。表9-9显示了突发强度持续到2018年的高突发性文献（Ziegler et al.，2009；Potts et al.，2012，2013；Li et al.，2007；Rahman et al.，2013；Zhan et al.，2012；Tang et al.，2010；Fox and Castella，2013；Wu et al.，2013；Xu et al.，2014；Sengloyluan et al.，2014）。从2013年、2014年、2015年和2016年开始被大量引用的文章分别有1篇、6篇、3篇和2篇。第1篇是东南亚地区迅速扩张的橡胶种植园替代现有土地对环境将造成的影响（Ziegler et al.，2009），从2014—2018年突发强度持续了5年，突发强度21.27；第2篇是胶乳共混法制备还原氧化石墨烯/天然橡胶纳米复合材料（Potts et al.，2012），突发强度24.74；第3篇是采用Landsat影像量化西双版纳土地利用/土地覆盖变化的研究（Li et al., 2006）；第4篇是构建橡胶树基因组序列框架图（Rahman et al.，2013）。这4篇文章在前面的聚类分析中也检测到了。第5篇、9篇和11篇文章涉及MODIS遥感数据和Rapideye影像光谱等识别橡胶林取代次生林和农田种植后的生态系景观格局变化（Li et al.，2012；Fox and Castella，2013；Xu et al., 2014）。第6篇、7篇和10篇涉及石墨烯或氧化石墨烯/天然橡胶纳米复合材料的制备工艺及其硫化动力学、导电性、透气性和机械强度的研究，是功能化石墨烯应用的一个研究进展（Zhan et al.，2012；Potts et al.，2013；Wu et al.，2013）。第8篇是橡胶树乳管蔗糖转运蛋白基因诱导表达的开创性研究（Tang et al.，2010）。第12篇是Sengloyluan等将环氧化天然橡胶（ENR）应用于以二氧化硅为增强填料的天然橡胶（NR）轮胎胎面配方中提升其性能（Sengloyluan et al.，2014），对ENR作为分散剂的应用有一定的指导意义。

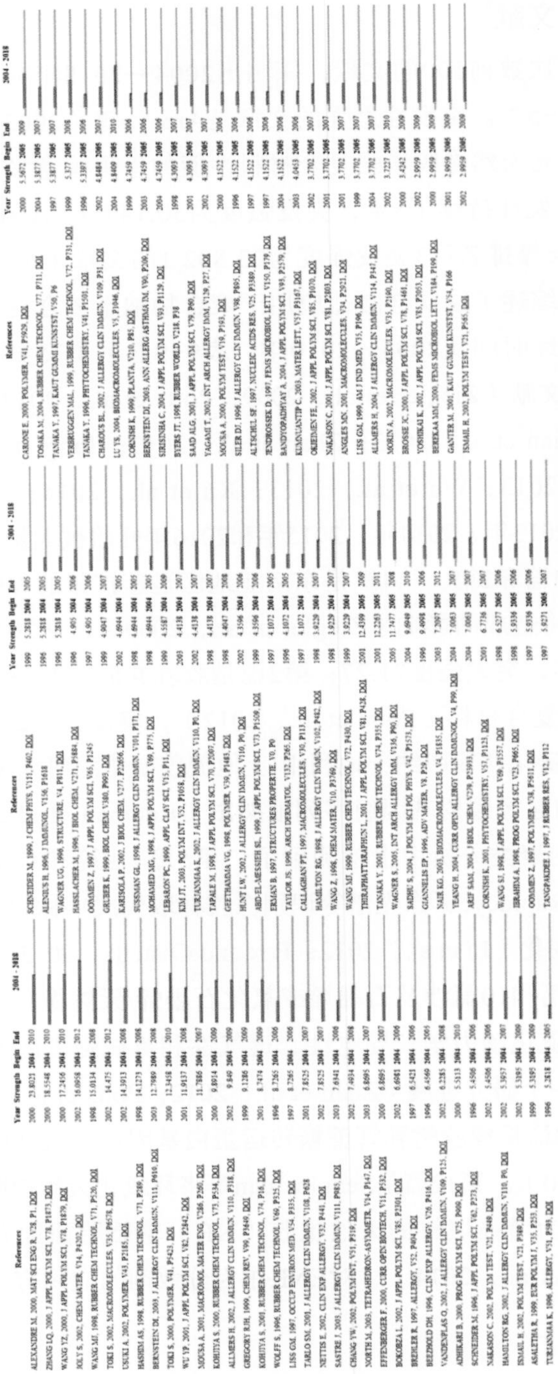

图9-5　2004—2018年引用突发强度排名前100的文献

Fig. 9-5　A total of 100 references with the strongest citation bursts during 2004–2018

表9–9 突发强度持续到2018年的文献

Tab. 9–9 References with the strongest citation bursts continuous 2018

作者（年） Author(Year)	标题 Title	引用次数 Cites	突发强度 Burst	持续时间 Duration	范围（2004—2018） Range（2004-2018）
Ziegler（2009）	The rubber juggernaut	109	21.27	2014—2018	
Potts（2012）	Processing-Morphology-Property relationships and composite theory analysis of reduced graphene oxide/natural rubber nanocomposites	91	24.74	2014—2018	
Li（2007）	Demand for rubber is causing the loss of high diversity rain forest in SW China	81	15.81	2013—2018	
Rahman（2013）	Draft genome sequence of the rubber tree *Hevea brasiliensis*	65	19.36	2014—2018	
Li（2012）	Mapping rubber tree growth in mainland Southeast Asia using Time-series MODIS 250m NDVI and statistical data	59	15.99	2014—2018	
Zhan（2012）	Enhancing electrical conductivity of rubber composites by constructing interconnected network of self-assembled graphene with latex mixing	56	15.17	2014—2018	
Potts（2013）	Latex and two-roll mill processing of thermally-exfoliated graphite oxide/natural rubber nanocomposites	51	13.81	2014—2018	
Tang（2010）	The sucrose transporter *HbSUT3* Plays an active role in sucrose loading to laticifer and rubber productivity in exploited trees of *Hevea brasiliensis*（para rubber tree）	47	13.65	2015—2018	
Fox（2013）	Expansion of rubber（*Hevea brasiliensis*）in mainland Southeast Asia：what are the prospects for smallholders?	44	15.47	2015—2018	
Wu（2013）	Vulcanization kinetics of graphene/natural rubber nanocomposites	43	15.12	2015—2018	
Xu（2014）	Landscape transformation through the use of ecological and socioeconomic indicators in Xishuangbanna，Southwest China，Mekong Region	36	15.15	2016—2018	
Sengloyluan（2014）	Silica-reinforced tire tread compounds compatibilized by using epoxidized natural rubber	29	15.14	2016—2018	

9.6 天然橡胶领域的新兴趋势

模块化度量了网络中的节点被划分为若干组的程度，从而使同一组中的节点比不同组之间的节点连接得更紧密。一个科学领域的集体知识结构可以表示为共同引用文献的关联网络。随着时间的推移网络发生演进变化，新发表的文章可能会引起网络结构上的变化，或者对结构变化影响很小，甚至没有影响（Chen et al.，2014；Chen，2017；Chen and Leydesdorff，2014；陈超美，2015；Garfield，2004）。

9.6.1 网络结构的变化

模块化度量了网络中的节点被划分为若干聚类的程度（Chen et al.，2014；Chen，2017；Chen and Leydesdorff，2014；陈超美，2015）。图9-6显示模块化和年发文量随时间的变化。发文量逐年增加，在2010年模块化出现大幅度下降。因此，2010年可能出现开创性的文章，也就是潜在的新兴趋势。如果2010发表的论文在后续出现引用突现，那么这些论文在改变整体知识结构方面发挥着重要作用。2010年有13篇论文出现引用突现（表9-10）（Tang et al.，2010；Valentín et al.，2010；Kim et al.，2010a，2010b；Guardiola-Claramonte et al.，2010；Duan et al.，2010；Valentín et al.，2010；Kuilla et al.，2010；Ma et al.，2010；Li et al.，2010；Bendahou et al.，2010；Peng et al.，2010；Wang et al.，2010），Tang等和Duan等的文章排在前列，这两篇文章代表了橡胶树乳管、树皮对乙烯利、茉莉酸甲酯和伤害等外界刺激诱导表达关键基因的重要研究（Tang et al.，2010；Duan et al.，2010），自2015年来出现引用激增。Valentin等是具有引文突现的最新文章之一，研究不同硫化体系对天然橡胶和聚丁二烯橡胶网络结构和动力学的定性差异（Valentín et al.，2010），自2016年突现强度激增。Guardiola-Claramonte等的文章代表了橡胶林取代原始森林为主的土地覆盖所带来的水文效应影响（Guardiola-Claramonte et al.，2010），自2014年开始引用激增。2010年发表了3篇综述，关于石墨烯、氧化石墨烯、改性石墨烯、碳纳米管和多壁碳纳米管等应用于弹性体或聚合物纳米复合材料制备及性能表征（Valentín et al.，2010；Kuilla et al.，2010；Ma et al.，2010）。表中其他文章讨论了不同填料补强天然橡胶纳米复合材料及性能的增强（Wu et al.，2013；Kim et al.，2010a，2010b；Li et al.，2010；Bendahou et al.，2010；Wang et al.，2010）。

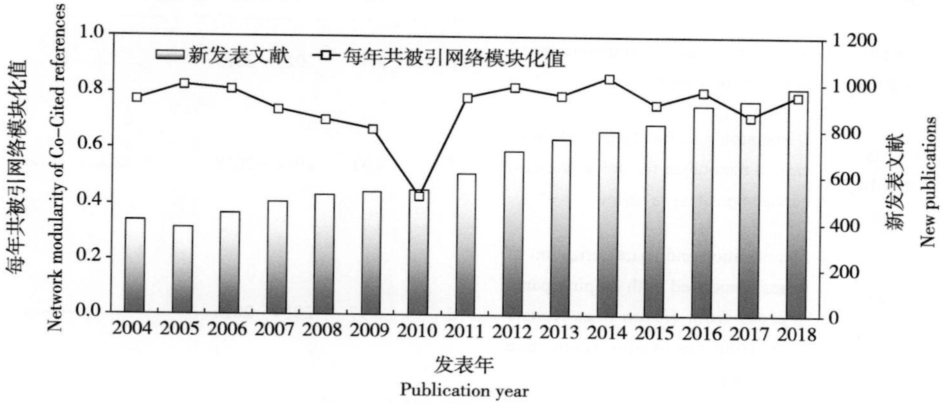

图9-6 2004—2018年间新发表的文献和网络的模块化程度

Fig. 9-6 New publications and modularity of the network during 2004-2018

表9-10 2010年发表具有引用突发的文献

Tab. 9-10 Articles published in 2010 with subsequent citation bursts

作者（年） Author（Year）	标题 Title	引用次数 Cites	突发强度 Burst	持续时间 Duration	范围（2004—2018） Range（2004-2018）
Tang（2010）	The sucrose transporter *HbSUT3* plays an active role in sucrose loading to laticifer and rubber productivity in exploited trees of *Hevea brasiliensis*（para rubber tree）	47	13.65	2015—2018	
Valentin（2010）	Inhomogeneities and chain dynamics in diene rubbers vulcanized with different cure systems	40	5.00	2016—2018	
Kim（2010a）	Graphene/polymer nanocomposites	36	8.75	2012—2015	
Guardiola-Claramonte（2010）	Hydrologic effects of the expansion of rubber（*Hevea brasiliensis*）in a tropical catchment	27	10.03	2014—2018	
Duan（2010）	Gene expression pattern in response to wounding, methyl jasmonate and ethylene in the bark of *Hevea brasiliensis*	27	9.48	2015—2018	
Valentin（2010）	Novel experimental approach to evaluate filler-elastomer interactions	22	8.00	2016—2018	

（续表）

作者（年） Author (Year)	标题 Title	引用次数 Cites	突发强度 Burst	持续时间 Duration	范围（2004—2018） Range（2004-2018）
Kuilla （2010）	Recent advances in graphene based polymer composites	21	3.44	2013—2015	
Ma （2010）	Dispersion and functionalization of carbon nanotubes for polymer-based nanocomposites：a review	17	6.31	2014—2018	
Li （2010）	Identification and characterization of genes associated with tapping panel dryness from *Hevea brasiliensis* latex using suppression subtractive hybridization	15	6.97	2014—2016	
Bendahou （2010）	Investigation on the effect of cellulosic nanoparticles' morphology on the properties of natural rubber based nanocomposites	15	6.92	2013—2015	
Peng （2010）	Self-assembled natural rubber/multi-walled carbon nanotube composites using latex compounding techniques	14	5.20	2014—2018	
Wang （2010）	Novel percolation phenomena and mechanism of strengthening elastomers by nanofillers	13	6.04	2014—2016	
Kim （2010b）	Graphene/polyurethane nanocomposites for improved gas barrier and electrical conductivity	13	6.00	2013—2015	

这些观察结果表明，2010年的模块化变化代表了天然橡胶研究的三大新趋势，一是橡胶树乳管、树皮关键基因表达；二是橡胶林取代热带雨林的土地利用变化下对生态效应影响；三是石墨烯、氧化石墨烯、碳纳米管等应用于天然橡胶纳米复合材料的合成及性能提升。这一趋势是当前活跃的，从2010年的引文激增数量就表明了这一趋势。

9.6.2 网络中的共被引文献聚类

时间线网络显示（图9-7），#2、#4和#7是最近形成的聚类，并且包含大量具有引文突发的红色圆环节点，因此将特别关注聚类#2、#4和#7，以确定天然橡胶领域的新兴趋势。最近发表的文章形成的最大聚类是#2巴西橡胶树，表9-11列出了聚类中引文突发强度最高的5篇文章（Rahman et al.，2013；Li et al.，2012；Ko et al.，2003；Triwitayakorn et al.，2011；Tang et al.，2010）。在这个聚类中突发性最强的文章Rahman等2013（Rahman et al.，2013），讨论了构建橡胶树基因组序列框架图并鉴定关键基因。该聚类的高突发性共被引文献中，共同的主题是通过转录组分析揭示橡胶树乳管、树皮中表达的关键基因。聚类#7是橡胶种植园，表9-12列出了突发强度最高的5篇文献（Ziegler et al.，2009；Li and Fox，2012；Li et al.，2007；Fox and Castella，2013；Xu et al.，2014）。在这5篇有代表性的文章中，一个共同的主题是热带雨林转变为橡胶林的土地利用变化下，对生态系统功能和经济发展等的影响。2012年，在土壤有机碳、生物量碳、生物多样性等生态学指标广泛应用的背景下，发表了聚类中突发强度较高的文章。

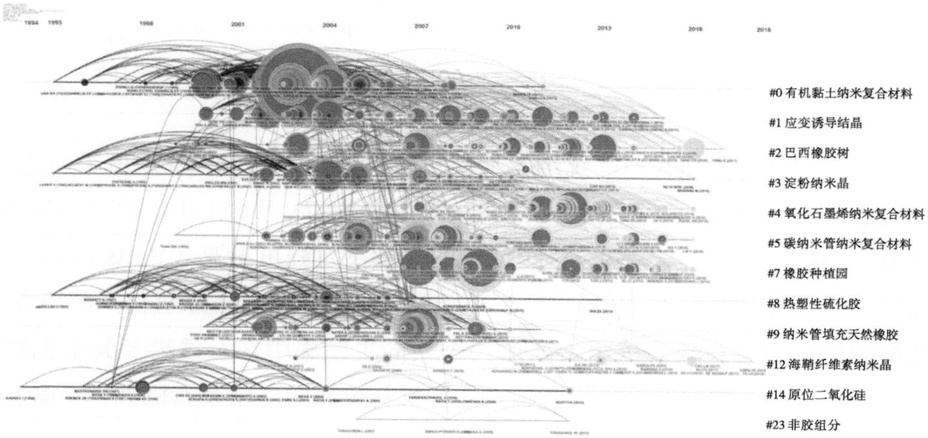

图9-7 共被引聚类的时间线网络

Fig. 9-7 Timelines of co-citation clusters

表9-11　聚类#2中的高突发性文献

Tab. 9-11　Articles with the strongest citation bursts in cluster #2

引用次数 Cites	突发强度 Burst	作者 Author	年 Year	标题 Title	期刊 Journal
65	19.36	Rahman	2013	Draft genome sequence of the rubber tree *Hevea brasiliensis*	BMC GENOMICS
50	19.17	Li	2012	De novo assembly and characterization of bark transcriptome using Illumina sequencing and development of EST-SSR markers in rubber tree（*Hevea brasiliensis* Muell. Arg.）	BMC GENOMICS
59	15.76	Ko	2003	Transcriptome analysis reveals novel features of the molecular events occurring in the laticifers of *Hevea brasiliensis*（para rubber tree）	PLANT MOL BIOL
41	15.70	Triwitayakorn	2011	Transcriptome sequencing of *Hevea brasiliensis* for development of microsatellite markers and construction of a Genetic Linkage Map	DNA RES
47	13.65	Tang	2010	Active role in sucrose loading to laticifer and rubber productivity in exploited trees of *Hevea brasiliensis*（para rubber tree）	PLANT CELL ENVIRON

聚类#4标记为氧化石墨烯纳米复合材料。表9-13显示突发强度最高的5篇文献（Potts et al.，2012，2013；Zhan et al.，2012；Wu et al.，2013；Zhan et al.，2011）。其中，Potts等2012年的文章突发强度最高（Potts et al.，2012），发表在Macromolecules上，关于胶乳共混法制备还原氧化石墨烯/天然橡胶纳米复合材料。这组文章的一个共同主题集中在石墨烯及其衍生物（氧化石墨烯）在制备天然橡胶纳米复合材料方面的应用及其机械、导电、传热、硫化动力学等性能的增强。

表9-12 聚类#4中的高突发性文献

Tab. 9-12 Articles with the strongest citation bursts in cluster #4

引用次数 Cites	突发强度 Burst	作者 Author	年 Year	标题 Title	期刊 Journal
109	21.27	Ziegler	2009	The rubber juggernaut	SCIENCE
59	15.99	Li	2012	Mapping rubber tree growth in mainland Southeast Asia using time-series MODIS 250m NDVI and statistical data	APPL GEOGR
81	15.81	Li	2007	Demand for rubber is causing the loss of high diversity rain forest in SW China	BIODIVERS CONSERV
44	15.47	Fox	2013	Expansion of rubber（*Hevea brasiliensis*）in Mainland Southeast Asia：what are the prospects for smallholders?	J PEASANT STUD
36	15.15	Xu	2014	Landscape transformation through the use of ecological and socioeconomic indicators in Xishuangbanna，Southwest China，Mekong Region	ECOL INDIC

表9-13 聚类#7中的高突发性文献

Tab. 9-13 Articles with the strongest citation bursts in cluster #7

引用次数 Cites	突发强度 Burst	作者 Author	年 Year	标题 Title	期刊 Journal
91	24.74	Potts	2012	Processing-Morphology-Property relationships and composite theory analysis of reduced graphene oxide/natural rubber nanocomposites	MACROMOLECULES
56	15.17	Zhan	2012	Enhancing electrical conductivity of rubber composites by constructing interconnected network of self-assembled graphene with latex mixing	J MATER CHEM
43	15.12	Wu	2013	Vulcanization kinetics of graphene/natural rubber nanocomposites	POLYMER
51	13.81	Potts	2013	Latex and two-roll mill processing of thermally-exfoliated graphite oxide/natural rubber nanocomposites	COMPOS SCI TECHNOL
70	13.40	Zhan	2011	Dispersion and exfoliation of graphene in rubber by an ultrasonically-assisted latex mixing and in situ reduction process	MACROMOL MATER ENG

9.7 本章小结

（1）天然橡胶是一个快速发展的研究领域，在共被引文献结构和时间模式分析的基础上，确定了3个主要的新兴趋势。3个新兴趋势都始于2010年，一是在诱导橡胶树乳管基因表达方面的开创性研究，包括随后的转录组分析用于橡胶树乳管、树皮表达关键基因鉴定；二是橡胶林取代热带雨林的土地利用变化下对生态、水文和经济效应等的影响；三是石墨烯或氧化石墨烯应用于天然橡胶纳米复合材料及其性能的提升。通过可视化分析概述了2004—2018年天然橡胶领域重要的里程碑意义文献。一些指标和观察结果表明，聚类#2在揭示橡胶树乳管、树皮中表达的关键基因方面发挥着关键和积极的作用；聚类#7关注热带雨林转变为橡胶林的土地利用变化下，对生态系统功能和经济发展等的影响；聚类#4侧重于石墨烯及其衍生物（氧化石墨烯）在制备天然橡胶纳米复合材料方面的应用研究。

（2）对学科类别、关键词和被引文献的突发性检测，检测到绿色可持续科技、遥感和地质学等学科的突发性都非常强劲，关键词“石墨烯”“动力学”“多样性”“种植园”“土地利用”等突发强度激增，预示着当前天然橡胶研究趋于多方向性。活跃的研究主题集中在两大学科领域：在天然橡胶加工领域，石墨烯和氧化石墨烯具有超高导电、超高导热、高效增强等优势特性，应用于高端天然橡胶纳米复合材料的研发；注重绿色、环保高端天然橡胶纳米复合材料的研发。在橡胶树植物科学领域，倾向于橡胶树产胶关键基因的分子生物学研究。随着遥感技术的发展和应用，新兴趋势还关注土地利用变化下橡胶林的生态、水文和经济效应等的研究。

（3）科学计量学的可视化分析技术通过追踪引文链接和引文突发之间的相互关系，并识别高度专业技术性的文章，能够引导研究人员关注一些最活跃和快速发展的前沿领域。分析中所识别的新兴趋势和模式是基于CiteSpace的选择计算特性，其设计旨在为科学前沿课题的构建提供相关领域的文献依据。关于相关主题综述文献数量的迅速增加，也是天然橡胶领域知识快速发展的一个标志。可视化分析工具可以在大量新发表的论文中发现重要研究进展。

（4）本章展示了一种科学计量方法，即利用该领域专家所发表的文献、信息和计算技术在不同抽象层次（被引文献和共被引文献聚类）上辨别模式和趋势，跟踪动态科学共同体的集体知识进展。希望使用的可视化分析工具能够在补充传统综述和调查文献方面发挥更积极的作用。

参考文献

陈超美，2015. 转折点创造性的本质[M]. 北京：科学出版社. 106–130.

国家天然橡胶产业技术体系，2016. 中国现代农业产业可持续发展战略研究天然橡胶分册[M]. 北京：中国农业出版社. 11–35.

Abdelmouleh M，Boufi S，Belgacem M N，et al.，2007. Short natural-fibre reinforced polyethylene and natural rubber composites：effect of silane coupling agents and fibres loading[J]. Composites Science and Technology，67（7-8）：1 627–1 639.

Angellier H，Molina-Boisseau S，Dufresne A，2005a. Mechanical properties of waxy maize starch nanocrystal reinforced natural rubber[J]. Macromolecules，38（22）：9 161–9 170.

Angellier H，Molina-Boisseau S，Lebrun L，et al.，2005b. Processing and structural properties of waxy maize starch nanocrystals reinforced natural rubber[J]. Macromolecules，38（9）：3 783–3 792.

Arroyo M，Lopez-Manchado M A，Herrero B，2003. Organo-montmorillonite as substitute of carbon black in natural rubber compounds[J]. Polymer，44（8）：2 447–2 453.

Bendahou A，Kaddami H，Dufresne A，2010. Investigation on the effect of cellulosic nanoparticles' morphology on the properties of natural rubber based nanocomposites[J]. European Polymer Journal，46（4）：609–620.

Berthelot K，Lecomte S，Estevez Y，et al.，2014. *Hevea brasiliensis* REF（Hev b1）and SRPP（Hev b3）：an overview on rubber particle proteins[J]. Biochimie，106：1–9.

Bhattacharyya S，Sinturel C，Bahloul O，et al.，2008. Improving reinforcement of natural rubber by networking of activated carbon nanotubes[J]. Carbon，46（7）：1 037–1 045.

Bhowmick A K，Bhattacharya M，Mitra S，2010. Exfoliation of nanolayer assemblies for improved natural rubber properties：methods and theory[J]. Journal of Elastomers and Plastics，42（6）：517–537.

Bokobza L，2007. Multiwall carbon nanotube elastomeric composites：a review[J]. Polymer，48（17）：4 907–4 920.

Bokobza L，2012. Enhanced electrical and mechanical properties of multiwall carbon nanotube rubber composites[J]. Polymers for Advanced Technologies，23（12）：1 543–1 549.

Börner K，Huang W，Linnemeier M，et al.，2010. Rete-netzwerk-red：analyzing and visualizing scholarly networks using the Network Workbench Tool[J]. Scientometrics，83（3）：863–876.

Carretero-González J，Retsos H，Verdejo R，et al.，2008. Effect of nanoclay on natural rubber microstructure[J]. Macromolecules，41（18）：6 763–6 772.

Chakraborty S，Kar S，Dasgupta S，et al.，2010a. Effect of treatment of Bis（3-triethoxysilyl propyl）tetrasulfane on physical property of *in situ* sodium activated and organomodified bentonite clay-SBR rubber nanocomposite[J]. Journal of Applied Polymer Science，116（3）：1 660–1 670.

Chakraborty S，Kar S，Dasgupta S，et al.，2010b. Study of the properties of in-situ sodium activated and organomodified bentonite clay-SBR rubber nanocomposites-Part I：characterization and rheometric properties[J]. Polymer Testing，29（2）：181–187.

Chen C M，2017. Science mapping：a systematic review of the literature[J]. Journal of Data and Information Science，2（2）：1–40.

Chen C M，Dubin R，Kim M C，2014. Emerging trends and new developments in regenerative medicine：a scientometric update（2000–2014）[J]. Expert Opinion on Bological Therapy，14（9）：1 295–1 317.

Chen C M，Leydesdorff L，2014. Patterns of connections and movements in dual-map overlays：A new method of publication portfolio analysis[J]. Journal of the Association for Information Science and Technology，65（2）：334–351.

Chen Y K，Xu C H，Wang Y P，2012. Viscoelasticity behaviors of lightly cured natural rubber/zinc dimethacrylate composites[J]. Polymer Composites，33（6）：967–975.

Chow K S，Wan K L，Isa M N M，et al.，2007. Insights into rubber biosynthesis from transcriptome analysis of *Hevea brasiliensis* latex[J]. Journal of Experimental Botany，58（10）：2 429–2 440.

Clough Y，Krishna V V，Corre M D，et al.，2016. Land-use choices follow profitability at the expense of ecological functions in Indonesian smallholder landscapes[J]. Nature Communications，7：13 137.

Das A，Stöckelhuber K W，Jurk R，et al.，2008. Modified and unmodified multiwalled carbon nanotubes in high performance solution-styrene–butadiene and butadiene rubber blends[J]. Polymer，49（24）：5 276–5 283.

Diani J，Fayolle B，Gilormini P，2009. A review on the Mullins effect[J]. European Polymer Journal，45（3）：601–612.

Duan C F，Argout X，Gébelin V，et al.，2013. Identification of the *Hevea brasiliensis* AP2/

ERF superfamily by RNA sequencing[J]. BMC Genomics, 14（1）: 30.

Duan C F, Rio M, Leclercq J, et al., 2010. Gene expression pattern in response to wounding, methyl jasmonate and ethylene in the bark of *Hevea brasiliensis*[J]. Tree Physiology, 30（10）: 1 349–1 359.

Dufresne A, 2010. Processing of polymer nanocomposites reinforced with polysaccharide nanocrystals[J]. Molecules, 15（6）: 4 111–4 128.

Fox J, Castella J C, 2013. Expansion of rubber（*Hevea brasiliensis*）in Mainland Southeast Asia: what are the prospects for smallholders? [J]. The Journal of Peasant Studies, 40（1）: 155–170.

Fröhlich J, Niedermeier W, Luginsland H D, 2005. The effect of filler–filler and filler–elastomer interaction on rubber reinforcement[J]. Composites Part A: Applied Science and Manufacturing, 36（4）: 449–460.

Garfield E, 2004. Historiographic mapping of knowledge domains literature[J]. Journal of Information Science, 30（2）: 119–145.

Ghasemi I, Karrabi M, Mohammadi M, et al., 2010. Evaluating the effect of processing conditions and organoclay content on the properties of styrene-butadiene rubber/organoclay nanocomposites by response surface methodology[J]. Express Polymer Letters, 4（2）: 62–70.

Golbon R, Cotter M, Sauerborn J, 2018. Climate change impact assessment on the potential rubber cultivating area in the Greater Mekong Subregion[J]. Environmental Research Letters, 13（8）: 084002.

Gopalan Nair K, Dufresne A, 2003. Crab shell chitin whisker reinforced natural rubber nanocomposites. 2. Mechanical behavior[J]. Biomacromolecules, 4（3）: 666–674.

Guardiola-Claramonte M, Troch P A, Ziegler A D, et al., 2010. Hydrologic effects of the expansion of rubber（*Hevea brasiliensis*）in a tropical catchment[J]. Ecohydrology, 3（3）: 306–314.

Guillaume T, Holtkamp A M, Damris M, et al., 2016a. Soil degradation in oil palm and rubber plantations under land resource scarcity[J]. Agriculture, Ecosystems & Environment, 232: 110-118.

Guillaume T, Maranguit D, Murtilaksono K, et al., 2016b. Sensitivity and resistance of soil fertility indicators to land-use changes: new concept and examples from conversion of Indonesian rainforest to plantations[J]. Ecological Indicators, 67: 49–57.

Hernández M, del Mar Bernal M, Verdejo R, et al., 2012. Overall performance of natural rubber/graphene nanocomposites[J]. Composites Science and Technology, 73: 40–46.

Huneau B，2011. Strain-induced crystallization of natural rubber：a review of X-ray diffraction investigations[J]. Rubber Chemistry and Technology，84（3）：425–452.

International Rubber Research and Development Board，2006. Portrait of the global rubber industry[M]. Kuala Lumpur：IRRDB，73–86.

Jacob M，Thomas S，Varughese K T，2004. Mechanical properties of sisal/oil palm hybrid fiber reinforced natural rubber composites[J]. Composites Science and Technology，64（7-8）：955–965.

Joly S，Garnaud G，Ollitrault R，et al.，2002. Organically modified layered silicates as reinforcing fillers for natural rubber[J]. Chemistry of Materials，14（10）：4 202–4 208.

Kapgate B P，Das C，Basu D，et al.，2014. Effect of silane integrated sol–gel derived in situ silica on the properties of nitrile rubber[J]. Journal of Applied Polymer Science，131（15）：40 531.

Kim H，Abdala A A，Macosko C W，2010a. Graphene/polymer nanocomposites[J]. Macromolecules，43（16）：6 515–6 530.

Kim H，Miura Y，Macosko C W，2010b. Graphene/polyurethane nanocomposites for improved gas barrier and electrical conductivity[J]. Chemistry of Materials，22（11）：3 441–3 450.

Kimura H，Dohi H，Kotani M，et al.，2010. Molecular dynamics and orientation of stretched rubber by solid-state ^{13}C NMR[J]. Polymer Journal，42（1）：25.

Ko J H，Chow K S，Han K H，2003. Transcriptome analysis reveals novel features of the molecular events occurring in the laticifers of *Hevea brasiliensis*（para rubber tree）[J]. Plant Molecular Biology，53（4）：479–492.

Kuilla T，Bhadra S，Yao D，et al.，2010. Recent advances in graphene based polymer composites[J]. Progress in Polymer Science，35（11）：1 350–1 375.

Li D J，Deng Z，Chen C L，et al.，2010. Identification and characterization of genes associated with tapping panel dryness from *Hevea brasiliensis* latex using suppression subtractive hybridization[J]. BMC Plant Biology，10（1）：140.

Li D J，Deng Z，Qin B，et al.，2012. De novo assembly and characterization of bark transcriptome using Illumina sequencing and development of EST-SSR markers in rubber tree（*Hevea brasiliensis* Muell. Arg.）[J]. BMC Genomics，13（1）：192.

Li H M，Aide T M，Ma Y X，et al.，2007. Demand for rubber is causing the loss of high diversity rain forest in SW China[J]. Biodiversity and Conservation，16（6）：1 731–1 745.

Li H M，Ma Y X，Aide T M，et al.，2008. Past，present and future land-use in Xishuangbanna，China and the implications for carbon dynamics[J]. Forest Ecology and

Management，255（1）：16–24.

Li Z，Fox J M，2012. Mapping rubber tree growth in mainland Southeast Asia using time-series MODIS 250m NDVI and statistical data[J]. Applied Geography，32（2）：420–432.

Lopez D，Amira M B，Brown D，et al.，2016. The *Hevea brasiliensis* XIP aquaporin subfamily：genomic，structural and functional characterizations with relevance to intensive latex harvesting[J]. Plant Molecular Biology，91（4-5）：375–396.

Ma P C，Siddiqui N A，Marom G，et al.，2010. Dispersion and functionalization of carbon nanotubes for polymer-based nanocomposites：a review[J]. Composites Part A：Applied Science and Manufacturing，41（10）：1 345–1 367.

Mensah B，Gupta K C，Kim H，et al.，2018. Graphene-reinforced elastomeric nanocomposites：a review[J]. Polymer Testing，68：160–184.

Nie Z Y，Kang G J，Duan C F，et al.，2016. Profiling ethylene-responsive genes expressed in the latex of the mature virgin rubber trees using cDNA microarray[J]. PloS One，11（3）：1–24.

Pal K，Rajasekar R，Pal S K，et al.，2010. Influence of fillers on NR/SBR blends containing ENR-organoclay nanocomposites：morphology and wear[J]. Journal of Nanoscience and Nanotechnology，10（5）：3 022–3 033.

Papageorgiou D G，Kinloch I A，Young R J，2015. Graphene/elastomer nanocomposites[J]. Carbon，95：460–484.

Pasquini D，de Morais Teixeira E，da Silva Curvelo A A，et al.，2010. Extraction of cellulose whiskers from cassava bagasse and their applications as reinforcing agent in natural rubber[J]. Industrial Crops and Products，32（3）：486–490.

Peng Z，Feng C F，Luo Y Y，et al.，2010. Self-assembled natural rubber/multi-walled carbon nanotube composites using latex compounding techniques[J]. Carbon，48（15）：4 497–4 503.

Potts J R，Shankar O，Du L，et al.，2012. Processing–morphology-property relationships and composite theory analysis of reduced graphene oxide/natural rubber nanocomposites[J]. Macromolecules，45（15）：6 045–6 055.

Potts J R，Shankar O，Murali S，et al.，2013. Latex and two-roll mill processing of thermally-exfoliated graphite oxide/natural rubber nanocomposites[J]. Composites Science and Technology，74：166–172.

Qiu J，2009. Where the rubber meets the garden[J]. Nature，457：246–247.

Rahman A Y A，Usharraj A O，Misra B B，et al.，2013. Draft genome sequence of the rubber tree *Hevea brasiliensis*[J]. BMC Genomics，14（1）：75.

Ray S S，Okamoto M，2003. Polymer/layered silicate nanocomposites：a review from preparation to processing[J]. Progress in Polymer Science，28（11）：1 539−1 641.

Rotolo D，Hicks D，Martin B R，et al.，2015. What is an emerging technology[J]. Research Policy，44（10）：1 827−1 843.

Sengloyluan K，Sahakaro K，Dierkes W K，et al.，2014. Silica-reinforced tire tread compounds compatibilized by using epoxidized natural rubber[J]. European Polymer Journal，51：69−79.

Shanmugharaj A M，Bae J H，Lee K Y，et al.，2007. Physical and chemical characteristics of multiwalled carbon nanotubes functionalized with aminosilane and its influence on the properties of natural rubber composites[J]. Composites Science and Technology，67（9）：1 813−1 822.

Sittiphan T，Prasassarakich P，Poompradub S，2014. Styrene grafted natural rubber reinforced by in situ silica generated via sol–gel technique[J]. Materials Science and Engineering：B，181：39−45.

Sriupayo J，Supaphol P，Blackwell J，et al.，2005a. Preparation and characterization of α-chitin whisker-reinforced poly（vinyl alcohol）nanocomposite films with or without heat treatment[J]. Polymer，46（15）：5 637−5 644.

Sriupayo J，Supaphol P，Blackwell J，et al.，2005b. Preparation and characterization of α-chitin whisker-reinforced chitosan nanocomposite films with or without heat treatment[J]. Carbohydrate Polymers，62（2）：130−136.

Srivastava S，Mishra Y，2018. Nanocarbon reinforced rubber nanocomposites：detailed insights about mechanical，dynamical mechanical properties，payne，and mullin effects[J]. Nanomaterials，8（11）：945.

Tang C R，Huang D B，Yang J G，et al.，2010. The sucrose transporter *HbSUT3* plays an active role in sucrose loading to laticifer and rubber productivity in exploited trees of *Hevea brasiliensis*（para rubber tree）[J]. Plant，Cell and Environment，33（10）：1 708−1 720.

Tang Z H，Wu X H，Guo B C，et al.，2012. Preparation of butadiene-styrene-vinyl pyridine rubber-graphene oxide hybrids through co-coagulation process and in situ interface tailoring[J]. Journal of Materials Chemistry，22（15）：7 492−7 501.

Teh P L，Ishak Z A M，Hashim A S，et al.，2004. Effects of epoxidized natural rubber as a compatibilizer in melt compounded natural rubber-organoclay nanocomposites[J]. European Polymer Journal，40（11）：2 513−2 521.

Toki S，Sics I，Ran S，et al.，2002. New insights into structural development in natural rubber during uniaxial deformation by in situ synchrotron X-ray diffraction[J]. Macromolecules，35（17）：6 578−6 584.

Tosaka M，Murakami S，Poompradub S，et al.，2004. Orientation and crystallization of natural rubber network as revealed by WAXD using synchrotron radiation[J]. Macromolecules，37（9）：3 299–3 309.

Triwitayakorn K，Chatkulkawin P，Kanjanawattanawong S，et al.，2011. Transcriptome sequencing of *Hevea brasiliensis* for development of microsatellite markers and construction of a genetic linkage map[J]. DNA Research，18（6）：471–482.

Valentín J L，Mora-Barrantes I，Carretero-González J，et al.，2010a. Novel experimental approach to evaluate filler-elastomer interactions[J]. Macromolecules，43（1）：334–346.

Valentín J L，Posadas P，Fernández-Torres A，et al.，2010b. Inhomogeneities and chain dynamics in diene rubbers vulcanized with different cure systems[J]. Macromolecules，43（9）：4 210–4 222.

van Beilen J B，Poirier Y，2007. Establishment of new crops for the production of natural rubber[J]. Trends in Biotechnology，25（11）：522–529.

van Eck N，Waltman L，2009. Software survey：VOSviewer，a computer program for bibliometric mapping[J]. Scientometrics，84（2）：523–538.

Varghese S，Karger-Kocsis J，2003. Natural rubber-based nanocomposites by latex compounding with layered silicates[J]. Polymer，44（17）：4 921–4 927.

Wang Z H，Liu J，Wu S Z，et al.，2010. Novel percolation phenomena and mechanism of strengthening elastomers by nanofillers[J]. Physical Chemistry Cemical Physics，12（12）：3 014–3 030.

Wei F，Luo S Q，Zheng Q K，et al.，2015. Transcriptome sequencing and comparative analysis reveal long-term flowing mechanisms in *Hevea brasiliensis* latex[J]. Gene，556（2）：153–162.

Weng G S，Huang G S，Qu L L，et al.，2010. Large-scale orientation in a vulcanized stretched natural rubber network：proved by in situ synchrotron X-ray diffraction characterization[J]. The Journal of Physical Chemistry B，114（21）：7 179–7 188.

Wigboldus S，Hammond J，Xu J，et al.，2017. Scaling green rubber cultivation in Southwest China-An integrative analysis of stakeholder perspectives[J]. Science of the Total Environment，580：1 475–1 482.

Wu S W，Tang Z H，Guo B C，et al.，2013. Effects of interfacial interaction on chain dynamics of rubber/graphene oxide hybrids：a dielectric relaxation spectroscopy study[J]. Rsc Advances，3（34）：14 549–14 559.

Wu J R，Xing W，Huang G S，et al.，2013. Vulcanization kinetics of graphene/natural rubber nanocomposites[J]. Polymer，54（13）：3 314–3 323.

Xu C H，Chen Y K，Huang J，et al.，2012a. Thermal aging on mechanical properties and crosslinked network of natural rubber/zinc Dimethacrylate composites[J]. Journal of Applied Polymer Science，124（3）：2 240–2 249.

Xu C H，Chen Y K，Zeng X G，2012b. A study on the crosslink network evolution of magnesium dimethacrylate/natural rubber composite[J]. Journal of Applied Polymer Science，125（3）：2 449–2 459.

Xu J C，Grumbine R E，Beckschäfer P，2014. Landscape transformation through the use of ecological and socioeconomic indicators in Xishuangbanna，Southwest China，Mekong Region[J]. Ecological Indicators，36：749–756.

Xu T W，Jia Z X，Wu L H，et al.，2017. Effect of acetone extract from natural rubber on the structure and interface interaction in NR/CB composites[J]. RSC Advances，7（42）：26 458–26 467.

Yi Z F，Wong G，Cannon C H，et al.，2014. Can carbon-trading schemes help to protect China's most diverse forest ecosystems? A case study from Xishuangbanna，Yunnan[J]. Land Use Policy，38：646–656.

Zeng M，Gao H N，Wu Y Q，et al.，2010. Preparation and characterization of nanocomposite films from chitin whisker and waterborne poly（ester-urethane）with or without ultra-sonification treatment[J]. Journal of Macromolecular Science，Part A：Pure and Applied Chemistry，47（8）：867–876.

Zhan Y，Lavorgna M，Buonocore G，et al.，2012. Enhancing electrical conductivity of rubber composites by constructing interconnected network of self-assembled graphene with latex mixing[J]. Journal of Materials Chemistry，22（21）：10 464–10 468.

Zhan Y H，Wu J K，Xia H S，et al.，2011. Dispersion and exfoliation of graphene in rubber by an ultrasonically-assisted latex mixing and in situ reduction process[J]. Macromolecular Materials and Engineering，296（7）：590–602.

Ziegler A D，Fox J M，Xu J，2009. The rubber juggernaut[J]. Science，324（5930）：1 024–1 025.

10 天然橡胶领域科研合作网络分析

10.1 引言

关于科学合作机理的研究一直是学术界共同关注的主题，科学合作是科研工作者为生产新的科学知识这一共同目的而在一起工作。科学合作已经成为科学研究中的一个重要方面，这源于科学发展的复杂性、技术的飞速变化、知识的动态增长以及高度发展的专业知识和技能，必须通过合作来解决复杂的科学研究问题（Hara et al.，2003）。世界范围内的知识流动日益频繁，国家间的科研合作日益成为国家间科技交流的重要途径，并催生出众多前沿领域的科研成果，创新、知识创造以及知识合作成为产业乃至整个区域经济可持续高效稳定增长的重要推动力（Scherngell and Barber，2011）。在合作的过程中，可以促进科学家之间的知识传播与交流，以及先进试验工具和试验设备的共享。同时，在科学合作的过程中，隐性知识和技艺也得到了转化和共享（Beaver and Rosen，1979）。文献计量学中通过共同发表论文来对科研合作进行测度。美国西北大学的Wuchty等对Web of Science中1955—2005年（51年）的1 990万篇文献数据和1975—2005年（31年）的210万份专利数据的分析表明，除了艺术与人文领域的论文合作者的平均数量保持稳定外，在科学与工程、社会科学领域以及专利的合作者的平均数量都呈上升趋势（Wuchty et al.，2007）。科研合作网络研究的对象主要包括专利合作网络（Breschi and Catalini，2010）、论文合作网络（Wal，2013）、著作合作网络（Hara et al.，2003；Scherngell and Hu，2011）以及研发合作网络（Balland，2012）等，其研究内容则聚焦于科研合作网络的结构特征及结构演化特征，并且倾向于确定网络行动者的中心性等（Cassi and Plunket，2015）。

本章基于共现分析等理论，借助信息可视化技术分析软件——VOSviewer，对国际和国内天然橡胶研究领域的文献进行共现网络分析，分别从国家/地区（宏观）、机构（中观）和作者（微观）层面展现天然橡胶研究领域科研合作的动态演化模式，通过分析国际间和国内天然橡胶研究领域的合作关系，对我国天然橡胶领域的合作研究以及天然橡胶产业优化提供决策支持。

10.2 数据来源与研究方法

10.2.1 数据来源

国际论文数据来源于科睿唯安的Web of Science™（简称WOS）核心合集的SCI-E（自然科学引文索引）和SSCI（社会科学引文索引）。国内论文数据来源于中国科学引文数据库（Chinese Science Citation Database，简称CSCD）。采用主题检索，检索策略如表10-1。由于科学文献是测度科研活动和科研产出的最有效的指标（赵学文和龚旭，2007），而国内外合著论文则可以反映科学研究国际和本土的合作特征，因此分析数据选取Article和Review两种文献类型。国际和国内论文发表时间跨度分别为2001—2020年和1989—2020年，检索时间为2020年6月27日。国际论文最终检索出13 375条文献及其所引用的13 243条引文文献数据，国内论文最终检索出2 699条文献及其所引用的442条引文文献数据。

表10–1 天然橡胶国际和国内文献数据检索策略

Tab. 10–1 Topic search queries used for data collection on international and domestic natural rubber papers

检索式 Retrieval type	记录数 Records	检索策略 Topic search
#1	11 012	主题：（“rubber tree*” or “*Hevea*” or “natur* rubber” or “natur* latex” or “nr latex” or “rubber latex” or “lactiferous plant” or “rubber yielding plant” or “rubber producing plant*” or “rubber producing crop*” or “rubber biosynthesis”）AND 文献类型：（Article OR Review）AND 语种：（English） 索引=SCI-EXPANDED，SSCI 时间跨度=2001-2020
#2	6 251	主题：（“rubber”）AND 主题：（“tapping” or “plantation*” or “yard” or “yield” or “productivity” or “garden” or “forest” or “NR” or “planting area*” or “planting” or “growing area*” or “growing state” or “intercrop*” or “in terplanting” or “latex production” or “latex drainage”）AND 文献类型：（Article OR Review）AND 语种：（English） 索引=SCI-EXPANDED，SSCI 时间跨度=2001-2020
#3	13 375	#1 OR #2 索引=SCI-EXPANDED，SSCI 时间跨度=2001-2020
#4	2 443	主题：（“橡胶树” or “巴西橡胶” or “天然橡胶” or “天然胶乳” or “橡胶胶乳” or “橡胶林” or “橡胶人工林” or “橡胶种植园” or “橡胶园” or “植胶区” or “橡胶生物合成”）AND 文献类型：（Article OR Review） 索引=CSCD 时间跨度=1989-2020检索语言=自动
#5	701	主题：（“橡胶”）AND 文献类型：（Article OR Review）AND 主题：（“产量” or “种植面积” or “种植” or “排胶” or “产胶” or “割胶” or “橡胶草” or “产胶植物” or “银胶菊” or “银色橡胶菊” or “杜仲胶” or “间作” or “套作” or “间种” or “套种”） 索引=CSCD 时间跨度=1989-2020检索语言=自动
#6	97	主题：（“rubber tree*” or “*Hevea*” or “natur* rubber” or “natur* latex” or “nr latex” or “rubber latex” or “lactiferous plant” or “rubber yielding plant” or “rubber producing plant*” or “rubber producing crop*” or “rubber biosynthesis”）AND 文献类型：（Article OR Review）AND 语种：（English） 索引=CSCD 时间跨度=1989-2020检索语言=自动
#7	74	主题：（“rubber”）AND 文献类型：（Article OR Review）AND 语种：（English）AND 主题：（“tapping” or “plantation*” or “yard” or “yield” or “productivity” or “garden” or “forest” or “NR” or “planting area*” or “planting” or “growing area*” or “growing state” or “intercrop*” or “int-erplanting” or “latex production” or “latex drainage”） 索引=CSCD 时间跨度=1989-2020检索语言=自动
#8	2 699	#4 OR #5 OR #6 OR #7 索引=CSCD 时间跨度=1989-2020检索语言=自动

10.2.2 研究方法

利用VOSviewer文本挖掘与可视化软件，从国家/地区合作的地理分布、国家/地区合作网络（宏观）、机构合作网络（中观）和作者合作网络（微观）等层面，对天然橡胶领域国际间和国内的科学合作网络进行可视化展示和分析。

10.3 国际天然橡胶研究科研合作分析

10.3.1 合作国家/地区的地理分布

以国家/地区为分析层次，分析了天然橡胶科学领域2001—2020年发表论文的地理分布情况，同时识别科研合作网络和时间演化过程。

每篇文章都根据在数据库中列出的作者地址被指定到一个国家/地区和一个机构。天然橡胶合作研究来自86个不同的国家/地区。图10-1表示国际天然橡胶科研合作国家/地区的地理分布，图中圆圈的大小代表合作发文量的多少，可以很直观地看出天然橡胶研究在全球的分布情况，大部分国家都位于亚洲和欧洲。在该领域作出贡献大小的国家依次是中国、美国、泰国、印度、马来西亚和法国等，这些国家为天然橡胶领域开创性的研究奠定了基础，对天然橡胶领域合作开展的研究较多。

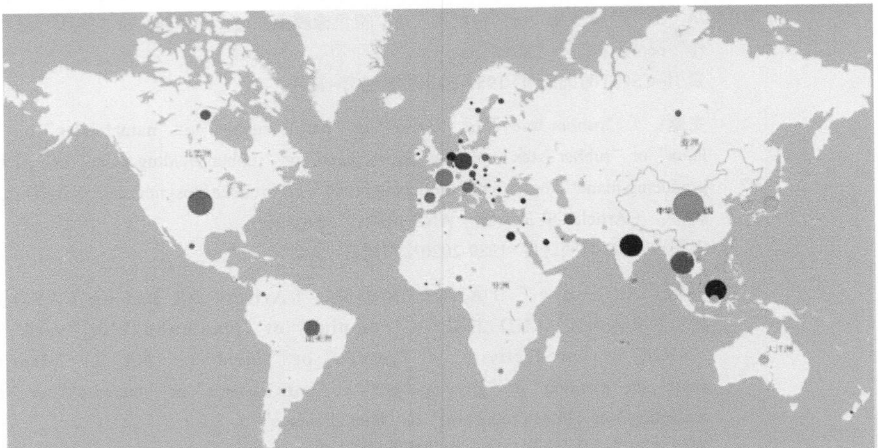

图10-1 国际天然橡胶科研合作国家/地区的地理分布

Fig. 10-1 The geographic distribution of collaboration countries/regions

10.3.2 国家/地区的合作网络

图10-2显示了天然橡胶研究中的国家/地区合作网络，其中，圆圈节点的大小表示发表论文的数量，连线的粗细表示合作的强度，节点和标签的大小与一个国家/地区的出现的次数成正比。根据这些国家/地区的合作强度，在合作网络中确定了六大群体，见图10-2（a）；每个国家/区域的平均出版年份见图10-2（b）。每个国家/地区的发文数量和引用次数等指标描述了在该研究领域高产国家的影响力（Dzikowski，2018），表10-2列出了国际天然橡胶研究合作中最高产的10个国家/地区，与图10-2中的网络相对应。

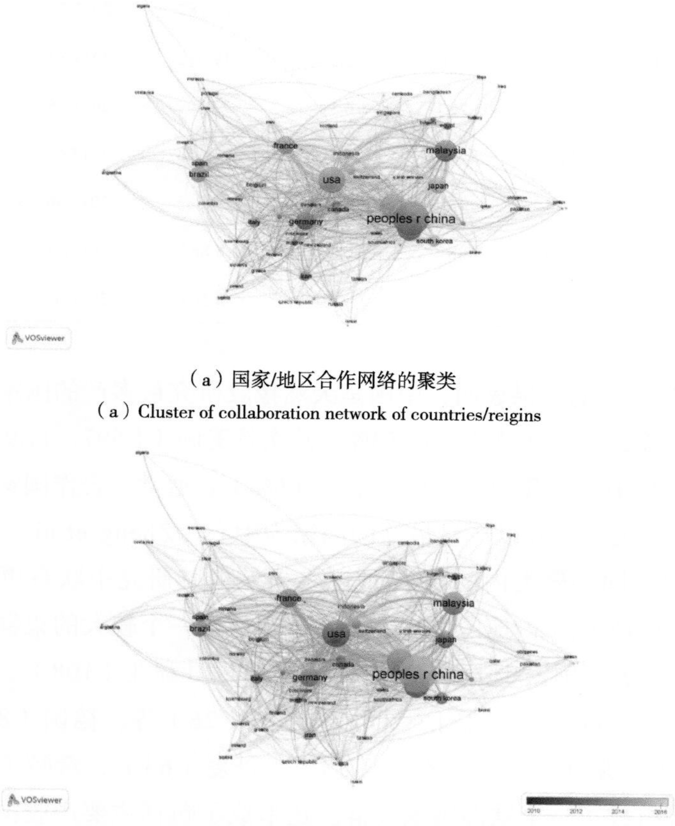

（a）国家/地区合作网络的聚类

（a）Cluster of collaboration network of countries/reigins

（b）国家/地区合作网络的平均出版年

（b）Average publication year of each country/region

图10–2　2001—2020年国际天然橡胶研究高产国家/地区合作网络

Fig. 10–2　Collaboration network of high productive countries/regions in international natural rubber research during 2001–2020

表10-2　国际天然研究领域合作发文前10位的国家/地区

Tab. 10-2　Top 10 high productive countries/regions in international natural rubber research

序号 No.	国家/地区 Countries/regions	区域 Continent	合作论文数 Cooperative documents	合作论文比例（%） Proportion（%）	引用次数 Total citations	平均出版年 Avg. pub. year	平均引用次数 Avg. citations
1	中国	亚洲	2 613	19.54	37 976	2014.75	14.53
2	美国	北美洲	1 597	11.94	41 660	2011.13	26.09
3	泰国	亚洲	1 493	11.16	20 384	2012.94	13.65
4	印度	亚洲	1 426	10.66	25 787	2011.20	18.08
5	马来西亚	亚洲	1 224	9.15	19 612	2012.33	16.02
6	法国	欧洲	875	6.54	27 899	2011.78	31.88
7	德国	欧洲	870	6.50	21 653	2012.43	24.89
8	日本	亚洲	745	5.57	13 885	2011.36	18.64
9	巴西	南美洲	688	5.14	8 862	2012.51	12.88
10	英国	欧洲	442	3.30	9 508	2011.52	21.51

图10-2（a）的结果表明，中国是天然橡胶研究最多产的国家，合作发表论文2 613篇，占论文总数的19.54%。其次是美国（1 597，11.94%）、泰国（1 493，11.16%）和印度（1 426，10.66%）。通常，合作国家往往在地理位置上相互关联，并以产量最高的国家为中心（Zheng et al.，2016）。在图10-2中，同一聚类内的国家/地区在天然橡胶研究中联系更为紧密。例如，中国与12个国家/地区进行了合作，组成了一个较大的集群，与中国合作排名前几名的国家/地区有英国（442）、尼日利亚（108）、斯里兰卡（97）、南非（64）、威尔士（31）和老挝（26）等。德国（870）、奥地利（136）、瑞典（92）、芬兰（88）、丹麦（63）、希腊（41）和挪威（27）等组成了一个欧洲国家集群。还形成了包括主要产胶国的几个集群：印度（1 426）、马来西亚（1 224）、荷兰（263）、埃及（215）、沙特阿拉伯（104）和土耳其（99）等；美国（1 597）、巴西（688）、意大利（336）、西班牙（325）、墨西哥（82）和哥伦比亚（47）等；日本

（745）、韩国（412）、澳大利亚（278）、印尼（225）、越南（94）和新加坡（69）等；法国（873）、比利时（104）、罗马尼亚（51）和葡萄牙（37）等；伊朗（310）、波兰（145）、俄罗斯（108）和中国台湾（76）等。如图10-2（b）所示，国家/地区的平均发表年份表明，目前，中国、泰国、马来西亚、巴西和德国是天然橡胶研究活跃的国家/地区。就平均引用次数而言，表10-2显示法国的平均贡献影响最高，其次是美国和德国。产出较高的国家，如中国和泰国，平均被引率相对较低。

橡胶工业的发展和发达国家备受瞩目的历史事件推动了天然橡胶研究的发展，例如美国1943年派出考察人员进入亚马孙平原找寻橡胶树树种。美国、法国、日本、英国等发达国家已经有了几十年成熟的橡胶工业。这些国家对于天然橡胶研究有着较长的历史，可以从较早的平均出版年份看出。另外，中国、泰国、马来西亚和巴西等发展中国家的天然橡胶产业近年来发展迅速。在这些发展中国家的天然橡胶产业中，技术创新变得越来越重要，天然橡胶研究在这些国家/地区得到了更多的关注和资金支持。

国际合作是在不同国家/地区之间转移知识和专业技能的一种方式。发展中国家/地区可以学习、利用和引进国外天然橡胶领域的研究方法和技术，从发达国家/地区获得关于技术、社会和组织进步的知识，并进行消化、吸收和国产化，以实现自主，改善产业发展状况。除了德国、奥地利、瑞典、芬兰、丹麦、希腊、挪威等欧洲发达国家之间的合作外，亚洲与欧洲、发展中国家/地区与发达国家/地区之间的国际合作网络也正在兴起。如中国、英国、尼日利亚、斯里兰卡、南非、威尔士和老挝之间的合作，美国、巴西、意大利、西班牙、墨西哥和哥伦比亚之间的合作，以及日本、韩国、澳大利亚、印尼、越南和新加坡之间的合作。

10.3.3 国际机构间的合作网络

图10-3显示了天然橡胶研究的关键机构之间的合作网络，包括来自86个国家和地区的806个院校。图10-3（a）显示了研究领域内机构的集群，而图10-3（b）展示了每个机构的平均出版年份聚类。表10-3列出了天然橡胶合作研究中最高产的10个机构。

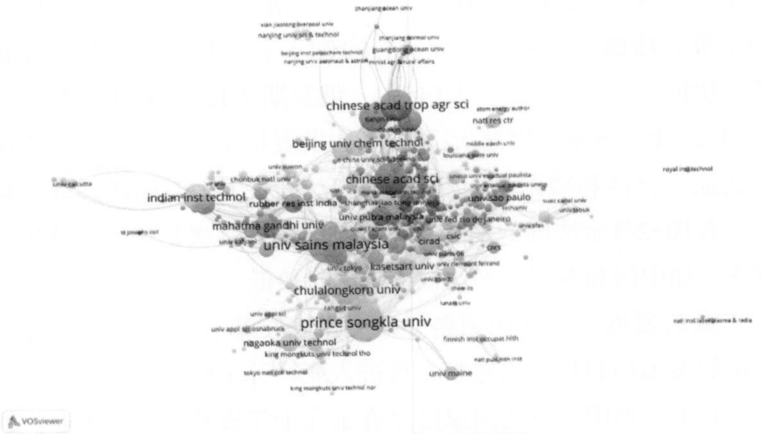

（a）研究机构合作网络的聚类

（a）Cluster of collaboration network of institutions

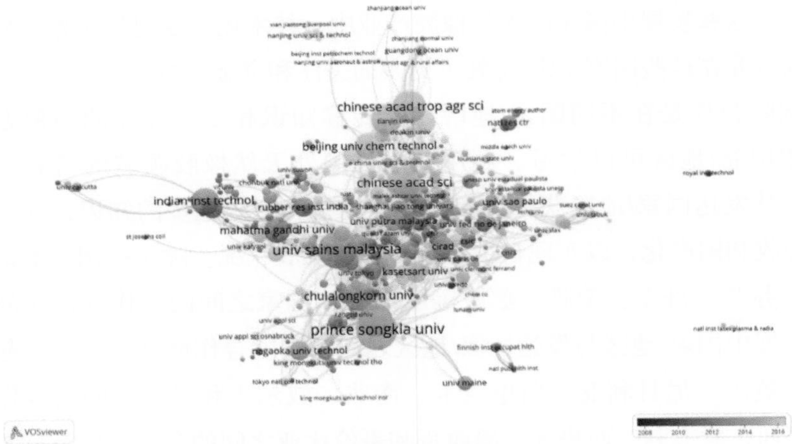

（b）研究机构合作网络的平均出版年

（b）Average publication year of each institution

图10–3　2001—2020年国际天然橡胶研究高产机构合作网络

Fig. 10–3　Collaboration network of high productive institutions in international natural rubber esearch during 2001–2020

表10-3 国际天然研究领域合作发文前10位的机构

Tab. 10-3 Top 10 high productive institutions in international natural rubber research

序号 No.	机构 Institutions	国家/地区 Countries/ regions	合作论文数 Cooperative documents	合作论文比例（%）Proportion（%）	引用次数 Total citations	平均出版年 Avg. pub. year	平均引用次数 Avg. citations
1	宋卡王子大学	泰国	507	3.79	6 064	2013.66	11.96
2	马来西亚理科大学	马来西亚	468	3.50	8 963	2011.62	19.15
3	中国科学院	中国	423	3.16	5 824	2014.63	16.88
4	中国热带农业科学院	中国	340	2.54	4 014	2014.35	11.81
5	玛希隆大学	泰国	336	2.51	6 256	2010.68	18.62
6	华南理工大学	中国	274	2.05	4 997	2014.66	18.24
7	朱拉隆功大学	泰国	270	2.02	4 148	2012.05	15.36
8	印度理工学院	印度	243	1.82	4 424	2010.42	18.21
9	北京化工大学	中国	218	1.63	3 976	2014.88	18.24
10	圣雄甘地大学	印度	204	1.53	6 963	2009.67	34.13

由表10-3可知，泰国宋卡王子大学共发表论文507篇，占全球论文总量的3.79%，在国际天然橡胶中排名第1。其次是马来西亚理科大学（468，3.50%）和中国科学院（423，3.16%），中国热带农业科学院排名第4（340，2.42%）。泰国宋卡王子大学是天然橡胶研究的主导机构，其科学与技术学院的橡胶技术与高分子材料系、生物塑料系致力于天然橡胶接枝改性、复合材料制备等研究。泰国宋卡王子大学共与34个机构进行了合作，其中玛希隆大学（336）、朱拉隆功大学（270）、泰国农业大学（137）、法国农业发展研究中心（135）、滑铁卢大学（71）合作的论文数量大于70篇。马来西亚理科大学的材料与矿产资源工程学院、工业技术学院等在天然橡胶复合材料性能增强方面作出了贡献，与37个科研院所进行了合作。马来西亚理科大学与马来西亚国立大学（180）、马来西亚橡胶部（113）、马来西亚大学（83）和泰诺马拉大学（81）合作论文数量大于70篇。中国科学院

西双版纳热带植物园围绕中国西南部橡胶人工林替代热带雨林导致生物多样性改变做了大量研究，中国科学院兰州化学物理研究所对天然橡胶复合材料的物理化学特性进行深入研究。中国科学院与69个科研机构进行了合作，与马来西亚博特拉大学（132）和美国农业部农业研究组织（61）合作论文数量大于70篇。就平均被引用次数而言，泰国圣雄甘地大学在前10名科研院所中具有最高的平均影响力。Ismail Hanafi和Nakason Charoen等领导了这些机构的天然橡胶研究工作。

为了更直观地可视化展示各机构的分布和论文的合作情况，各机构分布网络如图10-3（a）所示，可以清晰地观察到红色圆圈周围的合作网络。中国科学院位于网络的中心，与马来西亚博特拉大学、美国农业部农业研究组织、霍恩海姆大学等紧密合作。该网络还表明，来自同一国家/地区的机构通常比来自不同国家/地区的机构有更密切的合作关系。例如，中国的机构主要位于合作网络的中心和顶端，来自马来西亚、印度和巴西的机构位于网络的中部，而来自泰国的机构位于网络的下方。在该网络中，与其他院校相比，中国国科学院、复旦大学、泰国玛希隆大学、美国农业部农业研究组织、法国农业发展研究中心等具有更强的合作实力。从表10-3和图10-3（b）可以看出，中国的科研院所和高校对天然橡胶研究的重视程度越来越高，中国有4所院校排名前10。平均出版年表明，来自中国的机构（如中国科学院、中国热带农业科学院、华南理工大学、北京化工大学、海南大学、四川大学、青岛科技大学等）和亚洲的机构（泰国宋卡王子大学、泰国朱拉隆功大学、巴西圣保罗大学、马来西亚博特拉大学、马来西亚大学等）最近都活跃在该研究领域。

10.3.4 国际作者间的合作网络

作者的论文产出和合作关系是评价作者的定量指标。其中，作者的论文产出直接反映了作者在本研究领域的活跃程度，合作则反映了作者在学术共同体内部的社会关系。从数据源中提取作者出现频次等计量指标，通过发文数量和引用次数等指标描述在该研究领域高产作者的影响力（Dzikowski，2018）。表10-4列出了个人撰写或与他人合著超过7篇以上的前10位高产作者。

表10-4 国际天然橡胶研究合作论文数量超过7篇的前10位作者

Tab. 10-4 Top 10 authors published at least 7 papers in international natural rubber research

序号 No.	作者 Authors	合作论文数 Cooperative documents	引用次数 Total citations	平均出版年 Avg. pub. year	平均引用次数 Avg. citations
1	Ismail Hanafi	211	2 552	2013.20	12.09
2	Nakason Charoen	117	1 430	2013.84	12.22
3	Zhang Liqun	115	1 650	2016.51	14.35
4	Thomas Sabu	105	4 051	2012.14	38.58
5	Heinrich Gert	81	2 244	2013.85	27.70
6	Kawahara Seiichi	80	600	2013.20	7.50
7	Noordermeer Jacques W M	65	948	2012.98	14.58
8	Chen Yukun	61	1 329	2016.11	21.79
9	Riyajan Sa-ad	58	426	2014.07	7.34
10	Bhowmick Anil K	57	1 007	2011.56	17.67

通过VOSviewer将作者之间的合作关系可视化，如图10-4（a）和10-4（b）所示，协作关系更加明显。圆圈较大的高产作者（表10-4）通常与子网络（用不同颜色表示）中的其他作者联系关系较密切。数据显示，以Ismail Hanafi为合著最高产的作者，共211篇，其次是Nakason Charoen，共119篇，第三是Zhang Liqun（115），后面依次是Thomas Sabu（105）、Heinrich Gert（81）和Kawahara Seiichi（80）等。合著高产作者分别来自马来西亚理科大学、泰国宋卡王子大学、美国田纳西科技大学、泰国圣雄甘地大学、德国德累斯顿莱布尼茨固体与材料研究所和日本长冈技术科学大学。中国的另一位高产作者Chen Yukun来自华南理工大学。前10名研究者中最具影响力的作者是Thomas Sabu，平均每篇文章被引38.58次，其次是Heinrich Gert（27.70）和Chen Yukun（21.79）。相比之下，一些高产学者，如Zhang Liqun、Chen Yukun和Riyajan Sa-ad是近期较活跃的作者。

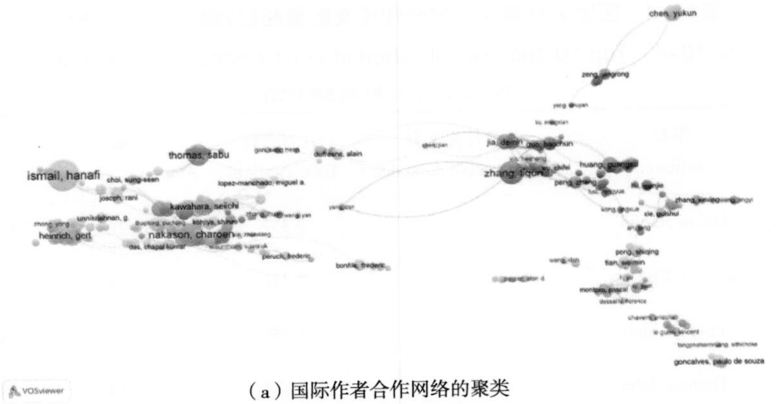

(a) 国际作者合作网络的聚类

(a) Cluster of collaboration network of international authors

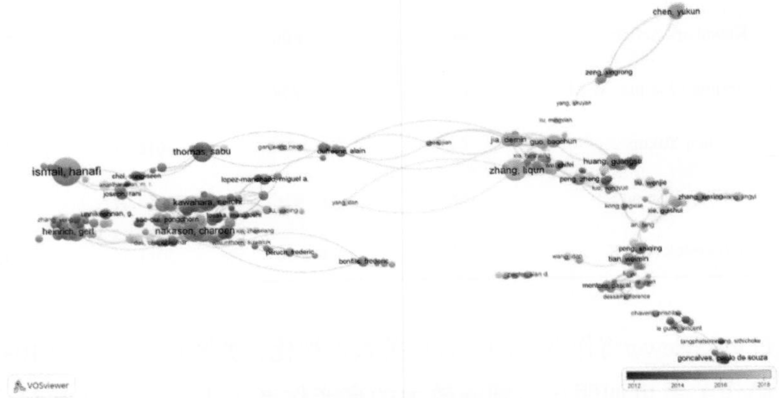

(b) 国际作者合作网络的平均出版年

(b) Average publication year of each international author

图10-4　2001—2020年国际天然橡胶研究高产作者合作网络

Fig. 10-4　Authors collaboration network of international natural rubber research during 2001–2020

10.4　国内天然橡胶研究科研合作分析

10.4.1　国内作者的合作网络

对国内天然橡胶研究发文量≥5篇的作者合作产出及合作网络进行分析，网络中共有406位作者，具体如图10-5（a）和图10-5（b）所示，高产作者的分布如表10-5所示。图中节点的大小反映了作者发表论文的数量，节点越大表示作者发表的论文越多。作者通过天然橡胶研究主题的论文合作，形成了不同的集群。集群内部作者合作关系密切，集群之间作者合作相对比较疏远。从国内天然橡胶学者合作的集群来看，合作网络主要以机构内部合作（团队型合作网络）和相近研究方向之间的合作（学科型合作网络）为主。由图10-5可看出，黄华孙（105，中国热带农业科学院橡胶研究所）、田维敏（73，中国热带农业科学院橡胶研究所）、李维国（63，中国热带农业科学院橡胶研究所）、谢贵水（13，中国热带农业科学院橡胶研究所）和廖双泉（52，海南大学材料与化工学院）发文量在50篇以上，是天然橡胶领域被CSCD期刊收录的高产作者。发文平均时间在2018年以后，且发文量不少于5篇的作者包括中国热带农业科学院橡胶研究所的薛欣欣（9），内蒙古民族大学生命科学学院的陈永胜（7）、黄凤兰（7）和李国瑞（6），云南省热带作物科学研究所的李小琴（7），中国热带农业科学院热带生物技术研究所的胡伟（7）和颜彦（6），海南大学热带农林学院的安邦（6）等，这些作者是近两年在天然橡胶研究领域活跃的高产学者。

在合作网络中，国内天然橡胶研究领域的合作作者共形成了25个集群。从分析结果来看，作者之间的合作以单位内部的合作为主，内部合作的核心成员是与其他单位合作的桥梁。其中，中国热带农业科学院橡胶研究所、中国热带农业科学院农产品加工研究所、华南理工大学、海南大学等在天然橡胶研究领域的合作表现突出，形成了若干个合作团队。如中国热带农业科学院橡胶研究所，形成了以黄华孙（发表论文105篇，合作人数16人）、罗微（发表论文43篇，合作人数26人）、李维国（发表论文63篇，合作人数17人）、田维敏（发表论文73篇，合作人数12人）等为核心成员的研究团队。反映了这些作者在网络中具有广泛的合作关系，并在各自的团队内部具有重要的作用。其中，最大的合作网络由43名作者构成，是华南理工大学材料科学与工程学院、中国热带农业科学院农产品加工研究所、广东海洋大学理学院的合作者们在天然橡胶加工领域的合作，最小的合作网络由5名作者组成。廖双泉、田维敏、李维国、唐朝荣是前10位合作者中近年来活跃在该研究领域的作者。

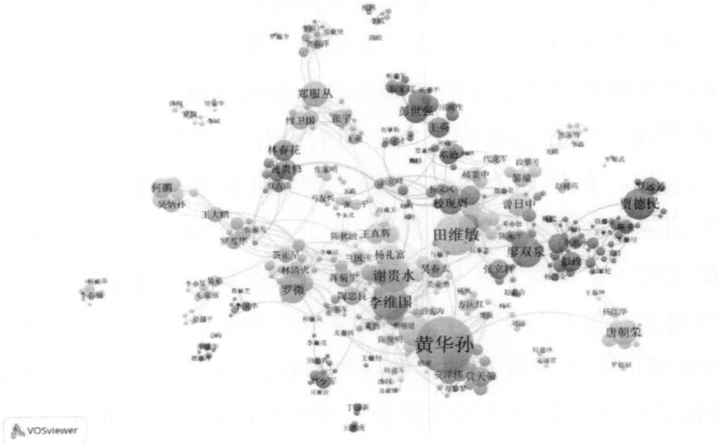

（a）国内作者合作网络的聚类

（a）Cluster of collaboration network of domestic authors

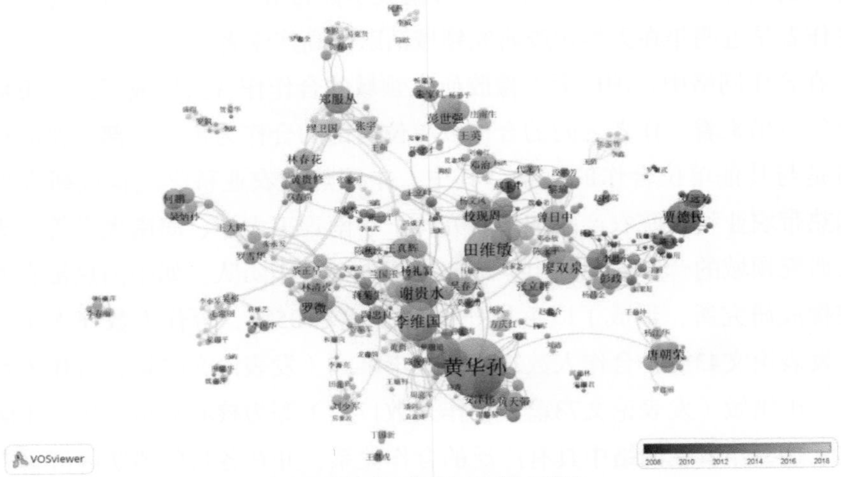

（b）国内作者合作网络的平均出版年

（b） Average publication year of each domestic author

图10-5　2001—2020年国内天然橡胶研究高产作者合作网络

Fig. 10-5　Authors collaboration network of domestic natural rubber research during 1989-2020

表10–5 国内天然橡胶研究合作论文数量超过5篇的前10位作者

Tab. 10–5 Top 10 authors published at least 5 papers in domestic natural rubber research

序号 No.	作者 Authors	合作论文数 Cooperative documents	平均出版年 Avg. pub. year
1	黄华孙	105	2010.98
2	田维敏	73	2013.29
3	李维国	63	2013.16
4	谢贵水	57	2011.26
5	贾德民	54	2006.70
6	林位夫	53	2009.91
7	廖双泉	52	2014.38
8	罗微	43	2011.91
9	郑服丛	43	2012.53
10	唐朝荣	41	2013.61

10.5 本章小结

本章以Web of Science核心集合和CSCD数据库的论文数据为基础，分别采集2001—2020年和1989—2020年天然橡胶科研论文数据，采用共现分析方法结合VOSviewer技术，研究国际间和国内天然橡胶研究的科研合作网络。主要结论如下：

（1）天然橡胶研究领域国家/地区的地理分布表明，亚洲的主要产胶国、欧洲和美洲的发达国家为该领域的研究奠定了基础。中国、美国、泰国、马来西亚、印度等是这些国家中产出率最高的。中国、泰国、马来西亚、巴西和德国等国家在这一研究领域作出了更大的贡献。

（2）就学术机构而言，泰国宋卡王子大学（泰国）是迄今为止文章数量最多的贡献者，马来西亚理科大学（马来西亚）、玛希隆大学（泰国）和中国科学院（中国）在论文数量、引用次数的平均影响力方面是领先的机构。近年来，包括中国科学院、中国热带农业科学院、华南理工大学和北京化工大学在内的一些中国科研机构和高校都非常活跃。来自泰国的宋卡王子大学和朱拉隆功大学近年来也很活跃。

（3）国际天然橡胶研究中论文产出表现突出的合作学者主要包括Ismail Hanafi、Nakason Charoen、Zhang Liqun、Thomas Sabu、Heinrich Gert和Kawahara Seiichi等。近期产出活跃的合作学者包括Zhang Liqun、Chen Yukun和Riyajan Sa-ad等。国内天然橡胶研究中论文产出表现突出的学者主要有黄华孙、田维敏、李维国、谢贵水和廖双泉，他们与网络中的作者建立了广泛的合作关系。廖双泉、田维敏、李维国和唐朝荣等是合作者中近年来活跃在该研究领域的作者。

（4）本章采用科研论文数据分析天然橡胶研究领域科研合作网络，其他例如专利合作、研发项目合作、专著合作等等都属于科研合作的范畴，因此今后的研究有必要深化对多形式的科研合作网络的结构进行综合分析，并且针对论文、专利等不同形式的科研合作网络的结构进行对比。运用论文合作数据目前只是体现在联系数量之上，而对于论文合作的领域今后也是研究的重点。

参考文献

赵学文，龚旭，2007. 科学研究绩效评估的理论与实践[M]. 北京：高等教育出版社. 77–83.

Balland P，2012. Proximity and the evolution of collaboration networks：evidence from research and development projects within the Global Navigation Satellite System（GNSS）Industry[J]. Regional Studies，46（6）：741–756.

Beaver D，Rosen R，1979. Studies in scientific collaboration[J]. Scientometrics，1（2）：133–149.

Breschi S，Catalini C，2010. Tracing the links between science and technology：an exploratory analysis of scientists' and inventors' networks[J]. Research Policy，39（1）：14–26.

Cassi L，Plunket A，2015. Research collaboration in co-inventor networks：combining closure，bridging and proximities[J]. Regional Studies，49（6）：936–954.

Dzikowski P，2018. A bibliometric analysis of born global firms[J]. Journal of Business Research，85：281–294.

Hara N，Solomon P，Kim S，et al.，2003. An emerging view of scientific collaboration：scientists' perspectives on collaboration and factors that impact collaboration[J]. Journal of the Association for Information Science and Technology，54（10）：952–965.

Scherngell T，Barber M J，2011. Distinct spatial characteristics of industrial and public research collaborations：evidence from the fifth EU Framework Programme[J]. Annals of Regional Science，46（2）：247–266.

Wal A L，2013. Cluster emergence and network evolution：a longitudinal analysis of the inventor network in sophia-antipolis[J]. Regional Studies，47（5）：651–668.

Wuchty S，Jones B F，Uzzi B，et al.，2007. The increasing dominance of teams in production of knowledge[J]. Science，316（5 827）：1 036–1 039.

Zheng T，Wang J，Wang Q，et al.，2016. A bibliometric analysis of micro/nano-bubble related research：current trends，present application，and future prospects[J]. Scientometrics，109（1）：53–71.